INTEGRATING PLANNING
WITH NATURE
Building climate resilience across the urban-to-rural gradient

ILLUSTRATED BY A CASE STUDY IN SAN JOSÉ AND COYOTE VALLEY

SFEI San Francisco Estuary Institute

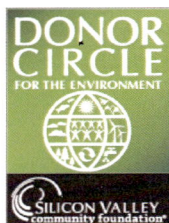

SFEI San Francisco Estuary Institute

SPUR

OPEN SPACE AUTHORITY SANTA CLARA VALLEY

DONOR CIRCLE FOR THE ENVIRONMENT — SILICON VALLEY community foundation

PREPARED BY SFEI

Authors

SFEI
Micaela Bazo
Matthew Benjamin
Erica Spotswood
Letitia Grenier

SPUR
Michelle Huttenhoff
Laura Tam
Laura Feinstein

OSA
Jake Smith
Marc Landgraf
Matt Freeman

Design & Production

SFEI
Ruth Askevold
Ellen Plane
Micaela Bazo
Matthew Benjamin

Model Places

SPUR
Benjamin Grant
Sarah Jo Szambelan

AECOM
Stephen Engblom
Cristian Bevington
Hugo Errazuriz

FUNDED BY the Silicon Valley Community Foundation's Donor Circle for the Environment and a charitable contribution from Google's Ecology Program

SFEI PUBLICATION 1013
November 2020

REPORT INFORMATION

This document was developed by the San Francisco Estuary Institute (SFEI) in partnership with the San Francisco Bay Area Planning and Urban Research Association (SPUR) and the Santa Clara Valley Open Space Authority (OSA). Integrated Planning with Nature is funded by the Silicon Valley Community Foundation (SVCF) Donor Circle for the Environment and a charitable contribution from Google's Ecology Program.

SUGGESTED CITATION

Bazo, M; Benjamin, M; Spotswood, E; Grenier, L; Huttenhoff, M; Tam, L; Feinstein, L; Smith, J; Landgraf, M; Freeman, M. 2020. Integrating Planning with Nature: Building climate resilience across the urban-to-rural gradient. SFEI Publication 1013, San Francisco Estuary Institute, Richmond, CA.

VERSION

v1.0 (November 2020)

REPORT AVAILABILITY

Report is available online at www.sfei.org

IMAGE PERMISSION

Permissions rights for images used in this publication have been specifically acquired for one-time use in this publication only. Further use or reproduction is prohibited without express written permission from the individual or institution credited. For permissions and reproductions inquiries, please contact the responsible source directly.

COVER IMAGE CREDITS

Top: Oblique view of San José. Photograph by Sergio Ruiz, courtesy of SPUR.
Bottom: Coyote Valley. Photograph by D. Neumann, courtesy of OSA.

CONTENTS

ACKNOWLEDGMENTS

Integrated Planning with Nature was developed in collaboration with the San Francisco Planning and Urban Research Association (SPUR) and Santa Clara Valley Open Space Authority (OSA). This project was funded by the Silicon Valley Community Foundation's Donor Circle for the Environment and a charitable contribution from Google's Ecology Program.

We would like to thank the group of experts that participated in our workshops or "charrettes" for their invaluable input:

- AECOM: Cristian Bevington, Diana Edwards, Stephen Engblom
- CA Dept of Conservation: Jeff Onsted, Kealii Bright
- CMG: Kevin Conger
- Committee for Green Foothills: Alice Kauffman
- ESA: Linda Peters, John Bourgeois
- Google: Kelly Garbach, Kate Randolph, Bethany Windle, Robin Bass
- Governor's Office of Planning and Research: Jennifer Phillips
- Greenbelt Alliance: Sarah Cardona
- Guadalupe Coyote Resource Conservation District: Stephanie Moreno
- HMH: Chris Telomen
- HT Harvey: Dan Stephens
- OSA: Andrea Mackenzie
- POST: Irina Kogan
- SAGE: Sibella Kraus
- SCC Office of Sustainability: Jasneet Sharma, Naresh Duggal
- SCC Planning and Development: Michael Meehan
- SCV Habitat Agency: Edmund Sullivan
- SJ Citywide Planning: Michael Brilliot, Jared Hart
- SJ Climate Smart Planning: Ed Schreiner
- SJ Department of Transportation: Jessica Zenk, Jim Ortbal
- SJ Environmental Services Department: Jeff Sinclair, Ken Davies, Kerrie Romanow
- SJ Parks, Recreation, and Neighborhood Services Department: Nicolle Burnham
- SJ Public Works: Michael O'Connell
- SPUR: Ben Grant
- Sitelab: Laura Crescimano
- TW Planning Consultants: Terrell Watt
- UC Small Farm Program: Aparna Gazula
- Valley Water: Brian Mendenhall, Samantha Greene

Special thanks to the *Model Places* team for allowing us to use their draft conceptual illustrations of future densification scenarios as templates for this report.

And thank you to Robin Grossinger (SFEI), Megan Wheeler (SFEI), and Joy Woo (AECOM) for their careful review and help with writing.

EXECUTIVE **SUMMARY**

The power of integrated planning with nature

Climate change and development pressures are creating an urgent need to build more resilience into the ecosystems we live in and rely upon. To create such resilience, planners need to couple rural and urban areas, because they are generally part of the same landscape system in which decisions made in one area affect outcomes in another. Silicon Valley exemplifies this situation, where land-use planning in Coyote Valley, higher in the watershed, affects outcomes in the city of San José, lower in the watershed, and vice versa. Such outcomes include flood risk, groundwater recharge, support for biodiversity, vehicle miles traveled (and the greenhouse gas consequences), and human health and well-being. For example, stormwater management in urban areas is far more effective if it is complemented by infiltration and flood detection upstream in rural landscapes.

◀ Left top: Bicycle in the cycling lane, Ist Street, San José. Photograph by Richard Masoner, courtesy of CC 2.0. Left bottom: California quail, Coyote Creek Valley. Photograph by Allan Hack, courtesy of CC 2.0

This project is a demonstration of how planning across the urban-to-rural gradient can create positive outcomes in both areas. This novel partnership brings together urban planning, from SPUR; rural planning, from OSA; and the best available science, from SFEI, to quantify the benefits that could be gained from integrated planning. This demonstration was placed in a realistic context by creating future scenarios with the input of stakeholders from San José and Coyote Valley through a series of planning workshops. Development pressure was included by assuming a more densely populated city in the future. The types of landscapes analyzed were drawn from San José and Coyote Valley.

The interventions that are recommended to improve the future landscape emphasize nature-based solutions. This focus is based on new research showing that natural

See page 94 for a comprehensive matrix of Policy and Planning Recommendations

Charleston Retention Basin and greenway in Mountain View, CA. Photograph by Shira Bezalel, SFEI.

infrastructure, like trees and wetlands, provides multiple benefits, is adaptive and resilient, and generally has low maintenance costs over time compared to traditional engineered approaches. For example, trees provide shade, carbon sequestration and storage, biodiversity support, and human health and well-being benefits. This project bridges the urban-to-rural divide to develop an approach that treats nature-based solutions as critical infrastructure central to adapting to the climate crisis and accommodating ongoing development.

Our findings are clear: development pressure can be accommodated in cities while improving the livability of urban areas, and without sprawling into rural areas. For example, in the Urban Neighborhood future scenario, the number of residents and jobs doubled, while greenspace area tripled. Tree canopy cover grew from 15 to 47%, which quadrupled carbon storage would significantly help mitigate rising temperatures and extreme heat events. In the Office Parks scenario, impervious cover was transformed into 37 new acres of greenspace for residents and workers, delivering more than five times the greenspace-per-capita minimum target set by the state of California. In the Cul-de-Sac Suburbs, tree plantings resulted in more than double the carbon sequestration, carbon storage, and avoided run-off, and nearly triple the amount of air pollutants removed, in combination with building housing for more than 1,000 residents. When aggregated, these nature-based interventions can help build regional climate resilience in alignment with addressing housing and affordability challenges, especially when coordinated with improvements in upstream rural areas.

The rural areas analysis showed that incorporating nature-based solutions in working and conserved lands could amplify ecosystem services, with benefits locally and for cities downstream. In the Parks & Protected Areas scenario, the restoration of wetlands and other natural areas had dual benefits of creating critical wildlife corridors while

functioning as regional stormwater infrastructure that benefits urban areas lower in the watershed. Furthermore, landowners collectively could get up to two million dollars in annual carbon offset payments for adopting climate-smart practices such as establishing hedgerows, mulching, applying compost, and restoring riparian buffers.

The strategies and solutions presented in this report vary widely in their implementation costs and associated benefits, though larger-scale actions generally reduce risks from floods, heat waves, droughts, and fires more effectively than more localized ones. Various policy tools exist to support these large-scale actions, and realizing them will require concerted efforts from multiple levels of government and diverse stakeholders.

This analysis demonstrates how integrating planning with nature can create healthy and beneficial landscapes. Implementing this approach in San José and Coyote Valley, as well as translating it to other geographies and scaling it up to larger areas, could derive significant benefits for people and wildlife, even as the climate changes and development pressures continue.

Mission Creek, San Francisco, at king tide. Photograph by Sergio Ruiz, courtesy of SPUR.

1
PLANNING WITH NATURE

Over the next century, the San Francisco Bay Area is poised to face three major challenges: adapting to a changing climate, adding infill development to accommodate a growing population, and maintaining natural and working lands in the face of development pressure. Our success as a region at resolving these challenges will depend on how we use our urban and rural land.

Left top: Looking down at the Viva Calle San José in 2017 (a celebration that temporarily closes miles of San José streets to bring communities together). Photograph by Sergio Ruiz, courtesy of SPUR. Left bottom: American coots in the tidal flats, Palo Alto Baylands. Photograph by Don DeBold, courtesy of CC 2.0.

To accommodate population growth while protecting open spaces, the Bay Area will need to increase density within existing urban footprints. Increasing density presents many benefits for climate change mitigation and human health. The concentration of people and industries in cities leads to greater innovation and creativity, economies of scale in infrastructure, and more efficient distribution of social services, education, and healthcare (Bettencourt et al. 2007; Sanderson et al. 2018). Higher density residential developments have lower energy use and greenhouse gas emissions per capita than lower density residential developments (Norman et al. 2006). Compact cities have higher road intersection density, greater diversity of land uses, more infrastructure network connectivity, and access to mass transit options, which reduce the need for driving and increase the likelihood of commuter and leisurely walking as well as overall levels of physical activity (Burton 2002; Ewing et al. 2003). This has implications not only for fuel-derived greenhouse gas emissions but also for mortality and health outcomes. Urban sprawl is associated with a greater prevalence of hypertension, obesity, traffic fatalities, pedestrian injuries, air pollution, and dangerously high ozone levels (Ewing et al. 2003; Ewing et al. 2003; Stone 2008). Given the long established health impacts of sedentary lifestyles and air pollution, it is no surprise that urban sprawl is a significant predictor of chronic medical conditions and lower health-related quality of life (Sturm and Cohen 2004).

Looking toward downtown San José Photograph by Sergio Ruiz, courtesy of SPUR.

Sprawling development patterns have also led to the large-scale conversion of natural areas and, consequently, to biodiversity loss. Rural areas, which have low population density and large swaths of undeveloped land, are essential for maintaining biodiversity: they provide critical habitat to sensitive species that do not tolerate urban conditions, facilitate regional wildlife movement (McDonald et al. 2020), and allow species to migrate beyond their traditional ranges to adapt to new climatic conditions (Pecl et al. 2017). Investing in conservation and smart land-use policies can ensure that rural areas contribute to regional resilience goals. Overall, developing rural areas results in a larger loss of ecosystem services (e.g., carbon storage, water infiltration, human well-being, agricultural production, pollination, pest control, noise reduction, air purification, and temperature regulation) compared to increasing density in suburban developments (Stott et al. 2015).

Whether we choose to sprawl or increase density is intimately connected to outcomes for both climate mitigation and adaptation. Sprawl increases the frequency and distance that people travel as well as the dependence on mostly gas-fueled passenger vehicles that drive up carbon emissions. Therefore, our ability to reduce emissions and sequester enough carbon to meet climate mitigation goals is strongly influenced by the extent of urban sprawl (Ewing et al. 2018). Climate adaptation will also be influenced by our choice to densify. Climate change adaptation depends on the ability of urban areas to tolerate novel climate patterns, including more frequent and extreme storms. Urban sprawl can increase flooding as well, especially when the conversion to impervious surfaces reduces infiltration upstream from cities (Lachman 2001).

Our ability to draw on nature-based solutions to mitigate and adapt to climate change and to deploy these solutions where they are most needed depends in part on whether we sprawl or densify. Whereas traditional approaches to climate adaptation rely on concrete and steel, nature-based solutions use techniques such as tree planting and marsh restoration to protect people from extreme heat and rising sea levels. Unlike engineered solutions, nature-based solutions can increase our capacity to adapt to a changing climate while providing many other benefits to people and ecosystems.

Urban sprawl leads to a loss of wildland area outside cities that could be used for both mitigation activities, such as tree planting or carbon farming to sequester carbon, and adaptation activities, such as restoring wetlands to increase groundwater recharge and reduce flooding in downstream urban areas. However, densification also comes with a significant risk; as infill development occurs, the loss of greenspace could lead to a loss in opportunities to deploy nature-based solutions within cities. Such interventions will be needed to make cities livable in the future. For example, the lack of greenspace in urban areas is associated with higher urban heat island intensities (Debbage and Shepherd 2015). Increasing urban density can lead to a loss of tree canopy and greenspace (Tratalos et al. 2007; Haaland and van den Bosch 2015). Therefore, a critical component to evaluating how best to reconcile density, open space protection outside cities, and climate adaptation is to consider where opportunities lie to increase and protect nature as cities densify.

Rural and urban planning are generally not integrated, and yet there are many benefits to planning at a system scale to create synergies and increase net benefits. Development rights in rural areas can be transferred to targeted growth areas in cities to reduce the loss of natural capital (i.e., the stock of natural resources that directly or indirectly provide goods and services to people) and facilitate the development of compact, walkable neighborhoods. Riparian, wetland, and floodplain restoration in rural areas can be part of an integrated strategy to build regional stormwater infrastructure and attenuate flooding in cities downstream. Increasing the efficiency of agricultural water use could help urban areas obtain their future water demand despite shifts in precipitation patterns and population growth (Flörke et al. 2018). Networks of ecological corridors in urban areas can help support regional biodiversity by providing habitat for locally endemic species (Freidin et al. 2011), facilitate migration (Seewagen et al. 2011), and help species adapt to climatic stress (Brans et al. 2017). Land use planning has profound implications on the configuration of ecological systems and the benefits they confer.

This report is the result of a unique partnership between a science nonprofit (SFEI), an urban planning think tank (SPUR), and a public open space agency (OSA). We see this project as a step towards planning across traditionally siloed sectors to generate interdisciplinary solutions to these interrelated challenges. **The primary goals of this report are to illustrate opportunities for using nature-based strategies in rural and densifying urban landscapes, quantify the benefits of these strategies, identify planning and policy approaches for implementation, and highlight how coordinating across the urban-rural divide can maximize the effectiveness of these measures.**

This report focuses specifically on nature-based solutions to climate change, which promise to reduce climate-related risks while providing additional benefits to people and nature. We use

San José and Coyote Valley as a case study to highlight the benefits of integrated planning. This project draws on current science and planning guidance, tools for quantifying climate benefits of nature-based solutions, and input from a large number of local experts who assisted in identifying local opportunities and constraints through participation in two workshops. The following sections offer further information on nature-based solutions, provide background on San José and Coyote Valley, describe the outcomes of workshops for each area, and summarize high-leverage nature-based solutions and policies that can be applied in this case study.

NATURE-BASED SOLUTIONS

Two forms of interventions exist for adapting to the effects of climate change: engineered and nature-based solutions. Engineered solutions typically rely on concrete and steel infrastructure, whereas nature-based solutions rely on restored natural or modified ecosystems. Many innovative hybrid approaches are now being developed that incorporate nature-based processes within partially engineered features. While engineered solutions are typically designed explicitly to protect people, nature-based and hybrid solutions simultaneously provide human well-being and biodiversity benefits (Cohen-Shacham et al. 2016).

Currently, municipalities often select engineered "gray" infrastructure solutions over nature-based "green" solutions, on grounds that the former are well understood and easier to permit. However, these approaches often provide fewer total benefits to communities and ecosystems, are expensive to maintain over time, give a false sense of security, and cannot adapt to changing conditions (Depietri and McPhearson 2017). For example, urban heat islands can be mitigated using shade structures and light-colored walkways, but these solutions have relatively limited capacity to cool the air, degrade over time, and provide few, if any, co-benefits (Akbari and Kolokotsa 2016). Meanwhile, trees reduce temperatures more than any other intervention, gain value as they grow, become more effective over time, and provide many additional benefits (e.g., capturing rainfall, sequestering carbon, capturing air pollutants, and supporting valued ecosystem functions) (Ong 2003). Unlike gray infrastructure, nature-based solutions can also help mitigate climate change by sequestering carbon and reducing emissions of greenhouse gases (Griscom et al. 2017). Quantifying the benefits of these nature-based solutions and planning for how and where to deploy them is a critical tool for motivating their use.

A varied toolbox of nature-based solutions applied in urban and rural areas has been shown to increase how well landscapes can provide desired functions and benefits as temperatures rise and severe storms become more frequent (Gago et al. 2013; Laurenson et al. 2013). Integrating across sectors that are usually managed independently will be key to making these solutions successful, given that, like any other system, ecosystems need to be managed holistically to maximize benefits. Similarly, planning for other critical climate mitigation and adaptation benefits at scales that encompass different land uses will enable us to optimize nature-based interventions, strategically placing them where they are most beneficial to the system as a whole.

Decision makers need actionable information that presents nature-based solutions in comparable terms with gray solutions to know which solutions are most likely to be effective at addressing urban sustainability challenges (Keeler et al. 2019). Ecosystem services, or the benefits people derive from nature, can be a useful tool for motivating cities to choose nature-based solutions over gray infrastructure (Costanza et al. 1997). Nature provides a long list of services, including heat reduction, air purification, and recreation. The valuation of ecosystem

☀ REFLECTIONS ON COVID-19

As we write this report, COVID-19 continues to ravage our communities and reshape our future in ways that we still don't fully understand.

This pandemic is undoubtedly a cataclysmic and tragic event but it can also be an opportunity for us to reflect on how to prepare for future global crises. Countries around the world have been forced to take drastic measures in an attempt to dampen infection rates. Keeping people from coming into close contact with each other has become crucial in preventing the transmission of this novel coronavirus. This has called into question the desirability of compact cities but has also led to swift shifts in public space dedication.

Many news media outlets have been quick to blame density for the rapid spread of COVID-19. Given the crisis is still unfolding and all of the ongoing challenges of systematic testing, there are not sufficient data available to ascertain whether people living in higher density urban areas are at greater risk. However, a preliminary analysis conducted by Drs. Robert McDonald and Erica Spotswood on data made

available by the New York Times suggests a weak relationship between density and spread of COVID-19 in the United States (R. McDonald and Spotswood 2020). High density areas registered cases earlier, but once the virus reached lower density communities it spread at a similar rate. This is consistent with NYU's analysis of trends in New York City, which suggests that higher rates of COVID-19 are not associated with overall population density but rather with overcrowding within units, as well as with communities of color who are less likely to be able to work from home (Furman Center 2020). Public health practices and infrastructure may prove to be more important in containing the spread and reducing mortality.

There is also a question about the viability of mass transit in the wake of this pandemic. Social distancing can be challenging on buses and trains, which were already running above their designed capacity before COVID-19. It is unclear when mass transit operators will be able to ease restrictions. But we must be careful not to revert to car-oriented spatial planning. People with fewer resources, a group that includes low-income communities and is growing as unemployment rates skyrocket, depend on mass transit to access job opportunities and meet day-to-day needs. Transit agencies will need to find creative solutions to adapt to funding gaps and public health protocols. In addition, promoting active mobility options, such as walking and biking, are still good investments that will have health-related benefits both during and after COVID-19. Planting trees along streets and building new greenways can help incentivize active mobility and make our streets more resilient and our cities better places to live.

In places like the Bay Area, where there has not been a full lockdown but the opportunities for recreation and exercise have been severely restricted, open space is in high demand. As the general manager of the East Bay Park Regional Park District, Bob Doyle, described, "visitation is insane. In my 45 years of park work, I've never seen these type of crowds, not ever" (Stienstra 2020). The City of Oakland closed seventy-five miles of city streets, or ten percent of its street network, to vehicular circulation to create more outdoor space for pedestrians and bicyclists during the shelter in place order (Bliss 2020). This 'slow streets' initiative is a testament to a city's ability to quickly adapt to changing circumstances, launching a previously controversial concept in less than a month. Seattle has gone even further, announcing that, given the success of its 'Stay Healthy Streets' program piloted during the pandemic, it will permanently close twenty miles of city streets to most vehicular traffic and accelerate the construction of bike infrastructure (Baruchman 2020). This should serve as a precedent for crisis responses of all kinds. However, criticism of the slow streets movement has brought to the surface the planning profession's ongoing failure to meaningfully include the public - especially communities of color - in planning processes, whether short-term or long. Addressing this shortcoming will be critical to making sure that quick, tactical responses to crises serve and protect everyone equitably.

This pandemic can be an opportunity to catalyze climate resilience planning. The climate crisis is still looming. Although emissions will go down this year as a result of quarantine measures, concentrations of greenhouse gas emissions keep increasing, and this year's reduction is still not enough to keep global temperatures from rising more than 1.5 degrees Celsius (Storrow 2020). The financial effects of the pandemic will also drastically impact city budgets in the coming years. In this challenging financial context, finding ways to maximize the multiple benefits of urban investments through integrated planning will be critical. Bold and drastic measures will be necessary to rechart our course and prepare us for the challenges that lay ahead.

Closed street with temporary play programming in response to COVID-19. Courtesy of Street Lab.

services allows people to compare the benefits of natural infrastructure to that of engineered solutions, and thus choose to maximize benefits in the face of limited resources. While quantifying ecosystem services always requires making some assumptions about current and future conditions, results can be integral for evaluating trade offs. The value of nature has generally been omitted in local land-use planning. While decision makers may not need ecosystem service cost–benefit assessments for nature-based solutions, knowing which approaches are most likely to succeed and when nature should be considered as an effective solution can help guide policy outcomes (Keeler et al. 2019). Demonstrating that ecological systems can confer more benefits than traditional gray infrastructure can encourage public and private actors to protect and restore nature. This report draws on ecosystem service quantification of nature-based solutions to illustrate their added value for urban and rural landscapes.

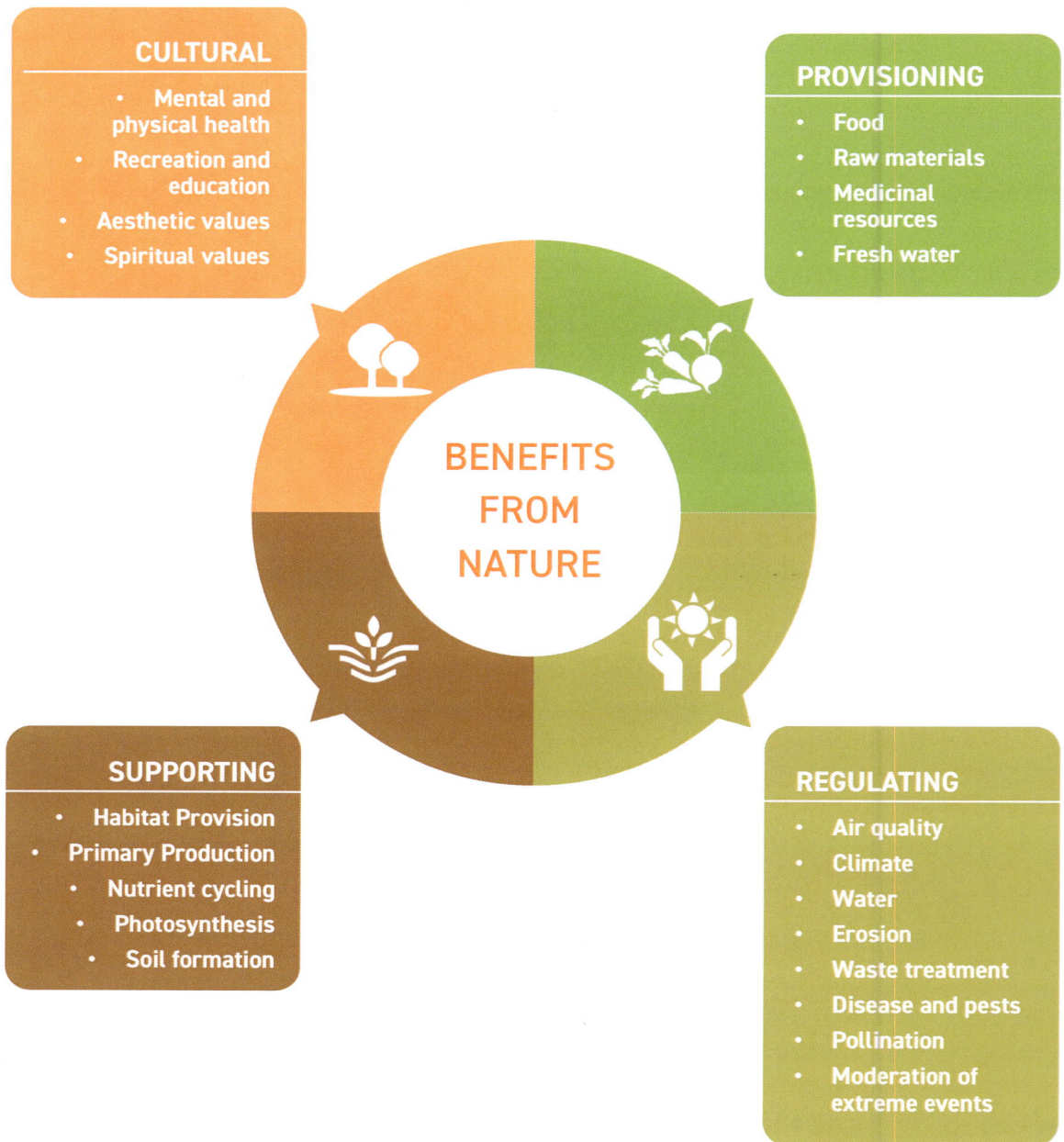

CULTURAL
- Mental and physical health
- Recreation and education
- Aesthetic values
- Spiritual values

PROVISIONING
- Food
- Raw materials
- Medicinal resources
- Fresh water

BENEFITS FROM NATURE

SUPPORTING
- Habitat Provision
- Primary Production
- Nutrient cycling
- Photosynthesis
- Soil formation

REGULATING
- Air quality
- Climate
- Water
- Erosion
- Waste treatment
- Disease and pests
- Pollination
- Moderation of extreme events

Summary of ecosystem services.

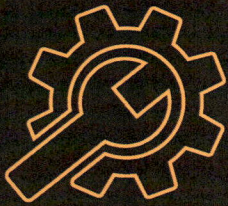

TOOLKIT OF NATURE-BASED SOLUTIONS

PARKS: Vegetated open spaces provide a wide variety of ecosystem services, such as reducing stormwater runoff, cooling the environment beyond their boundaries, and offering recreational opportunities. Larger, connected greenspaces densely planted with native vegetation and high tree canopy cover deliver more benefits to people and biodiversity.

RIPARIAN CORRIDORS AND BUFFERS: Riparian corridors, especially those with complex native vegetation, provide invaluable aquatic and terrestrial habitat in urban and agricultural areas, as well as regional habitat connections. Restoring channel profiles, widening riparian corridors, increasing setbacks, and expanding floodplains can greatly reduce the risk of flooding and improve water quality while providing opportunities for recreation and climate regulation.

GREENWAYS: Linear, vegetated open spaces provide a wide array of ecological, health, social, and economic benefits. Greenways provide space for bicycling and walking, and their presence can reduce reliance on automobiles. Vegetation in greenways can also help to reduce urban heat, slow runoff, remove air pollutants, buffer noise, and serve as valuable wildlife corridors in the urban landscape.

Plaza Cesar Chavéz Photograph by Sergio Ruiz, courtesy of SPUR.

BIORETENTION SYSTEMS: Planted stormwater detention and retention areas capture runoff and prevent pollutants from entering waterways. Bioretention features can take a variety of forms, from swales along streets to large retention basins within greenspaces. This type of hybrid infrastructure combines natural and engineered solutions to regulate flooding, increase groundwater recharge, and improve water quality.

FRONT AND BACK YARD IMPROVEMENTS: Yards with large, irrigated lawns require significant chemical, energy, and water inputs. However, by reducing the area of irrigated lawns, planting locally native, water-wise vegetation, and placing trees near buildings, these spaces can better support native biodiversity and contribute to climate resilience. Planting native vegetation in yards can help create ecological stepping stones, or habitat patches when aggregated, for urban wildlife.

URBAN FOREST: Trees sequester and store carbon, cool the environment, reduce air pollution, capture and store rainfall, facilitate water infiltration, and increase habitat connectivity. They can be integrated into many of the nature-based solutions listed both on public and private land. Rows of trees planted at regular intervals along streets improve outdoor thermal comfort and promote active mobility. A coordinated urban forest strategy can help improve local microclimates and create extensive, decentralized ecological networks that distribute ecosystem services to more people.

GREEN ROOFS: Roof surfaces either partially or completely covered with vegetation can reduce stormwater runoff, improve building energy efficiency, and create pollinator habitat. Benefits vary between intensive and extensive systems. Intensive green roofs have deeper soil profiles and can support larger plants or trees, hold more water, and offer greater thermal insulation. Green roofs on lower buildings are more beneficial to pollinators.

GREEN TERRACES: Balconies and courtyards provide additional opportunities for greening. They provide many of the same benefits as green roofs, but generally at a smaller scale. Green terraces with trees and deeper soil profiles are more effective at reducing runoff, cooling the environment, and sequestering and storing carbon, compared to unvegetated terraces. Such features can create ecological ladders that connect to green roof systems and offer biophilic benefits to people indoors.

GREEN WALLS: Green walls improve building energy efficiency and reduce air pollution, particularly when applied to the exterior of buildings in narrow street canyons. Trellis systems provide many of the benefits of green walls while requiring less complicated implementation or maintenance.

HEDGEROWS: Densely vegetated rows of woody plants or other perennial plants can provide valuable habitat and linkages through farmland. Hedgerows can be used to establish pollinator corridors, intercept particulate matter, increase carbon storage, and provide substrate for beneficial invertebrates as part of an integrated pest management strategy (NRCS 2012).

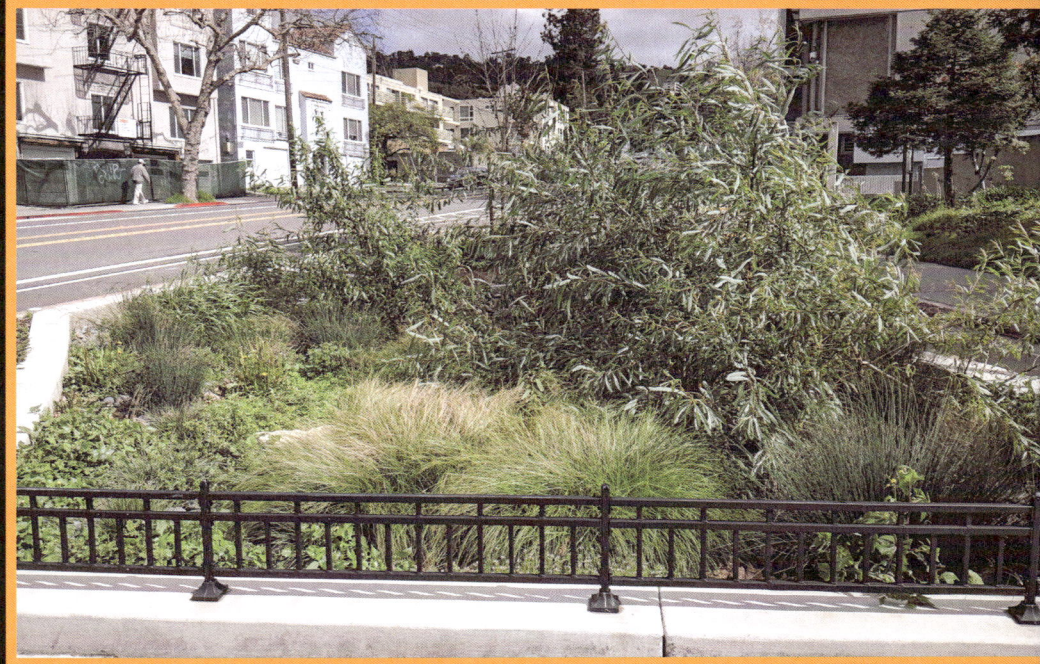

Stormwater detention basin with native plants in Berkeley, CA. Photograph by Robin Grossinger, SFEI.

Bosco Verticle in Milan, Italy , 2014. Photograph by Lorenzoclick, courtesy of CC 2.0.

RE-OAKING: Historically vast oak woodland ecosystems extended throughout the Bay Area. Restoring native oak trees and associated species in cities, farmland, and natural areas can benefit people and biodiversity. Oaks are drought-tolerant, have broad and dense canopies that mitigate heat stress, sequester more carbon than many of the most commonly used tree species, and support native wildlife (Spotswood et al. 2017).

WETLAND RESTORATION: Restoring seasonal and perennial wetlands can improve water quality, buffer flooding downstream, replenish groundwater, sequester carbon, fuel the aquatic food web, and provide sometimes regionally rare habitat for resident and migratory wildlife.

COVER CROPS: Planting grasses, legumes, forbs, and other groundcover crops between rows or underneath orchards can provide seasonal cover and other conservation benefits. Cover crops reduce erosion from wind and water, increase biodiversity, regulate soil moisture, and improve the soil's ability to store and sequester carbon (NRCS 2011).

COMPOST APPLICATION: Applying compost to croplands and rangelands improves soil health, increasing soil microbial organisms and plant biomass, and therefore carbon sequestration and storage. Compost application also improves the soil's ability to retain water and nutrients, increases water infiltration, and reduces erosion (Gravuer 2016). Compost in grasslands should only be applied in areas where it will not impact native species.

REDUCED TILLING: Limiting soil disturbance decreases tillage-induced dust particulate emissions that decrease air quality. Reducing or even limiting tilling can also lessen erosion, maintain or improve soil health and quality, increase plant-available moisture, and reduce fossil fuel-derived energy use (NRCS 2014).

MULCHING: Applying mulch in agriculture and gardens reduces erosion, protects soil from compaction, improves moisture retention, and suppresses weeds. This can in turn reduce irrigation and use of chemical inputs in farmland, as well as improve soil and plant health which increases carbon sequestration (Sharath et al. 2019).

PRESCRIBED GRAZING: Managing the intensity, frequency, timing, duration, and distribution of grazing can help achieve ecological objectives. Prescribed grazing can reduce erosion, control invasive plant species, and manage fire fuel loads (NRCS 2017).

Cover crops between rows of grape vines. Photograph by Stefano Lubiana, courtesy of CC 2.0.

Riparian buffers in agricultural fields south of Dallas, Texas. Courtesy of National Agroforestry Center.

2
CLIMATE RISK
in San José and Coyote Valley

In this report, we use San José and Coyote Valley as a case study for how natural systems can be protected and restored in urban and rural landscapes. This chapter includes background information on the study area, outlines major climate-related risks (i.e., fluvial flooding, drought, rising temperatures, wildfire, air pollution, reduced agricultural productivity, and environmental injustice), and identifies potential adaptation solutions.

◀ Left top: Coyote Creek after a rainstorm, February 2017. Photograph by Don DeBold, courtesy of CC 2.0. Left bottom: Putting out a fire near Scotts Valley, CA during the 2020 CZU Lightning Complex wildfires. Photograph by California National Guard.

San José

The City of San José sits at the southern end of the San Francisco Bay, at the heart of what is colloquially known as Silicon Valley. It is the tenth most populous city in the United States, and the largest in the Bay Area, with an estimated 1 million people in 2018 (US Census Bureau 2018) and projected growth to 1.3 million by 2050 (Romanow et al. 2018). San José has shown a commitment to sustainability and is leading efforts to respond to climate change. The City was able to reduce its water consumption by 28% during a recent drought, has the highest number of electric vehicles per capita in the United States, and has set an ambitious plan for decarbonization (Romanow et al. 2018). There are a number of plans underway in the City to integrate progressive climate actions, making San José a ripe example for integrative landscape planning.

The City encapsulates a variety of land uses, from a dense downtown area to dispersed office parks and undeveloped natural areas. Cul-de-Sac Suburbs, however, compose the majority of San José's sprawling landscape. A total of 94% of the City is zoned for single-family residential housing, compared to 37% in San Francisco and 75% in Los Angeles (Lopez 2019). The resulting sprawl makes these areas difficult to service with public transit, and has led to the construction of northern California's largest municipal road network, of about 2,400 miles, to support single-passenger vehicles (City of San José 2015).

> "We don't have a lot of vacant land left. But we need to densify. We need to transform the suburban city into a more urban place."
>
> _ Michael Brilliot, Deputy Director of City Planning (Lopez 2019)

San José sits at the interface of three unique and expansive habitat areas. The Santa Cruz Mountains in the southwest receive relatively high levels of rainfall and support hundreds of acres of redwood forests and oak woodlands, among other habitat types. In the Diablo Range to the east, relatively drier conditions support ecosystems such as chaparral, grasslands, and oak savanna (Bay Area Open Space Council 2019). To the north, the City's shoreline boasts extensive tidal marshes that provide habitat for numerous migratory waterbirds, endemic marsh wildlife, and aquatic species (Bay Area Open Space Council 2019).

Prior to development and sweeping modifications to the City's landscape, San José was a key crossroads between these three habitat areas. Upland parts of historical San José largely were oak woodlands, whereas areas with higher groundwater near the shoreline and along creeks fostered wet meadows, willow groves, and riparian forests.

Today, however, habitat areas within the City's boundaries are highly fragmented. City parks and other green spaces dot the landscape, but are generally separated by highways, housing developments, and other areas of low canopy cover. Streams and their adjacent wooded riparian corridors act as the primary connections between habitat patches in the City and the wildlands on the City's periphery.

SAN JOSÉ

Coyote Creek

COYOTE VALLEY

N

Figure 1: Aerial image of study area which includes urban areas in the City of San José and rural land in Coyote Valley.

5 miles

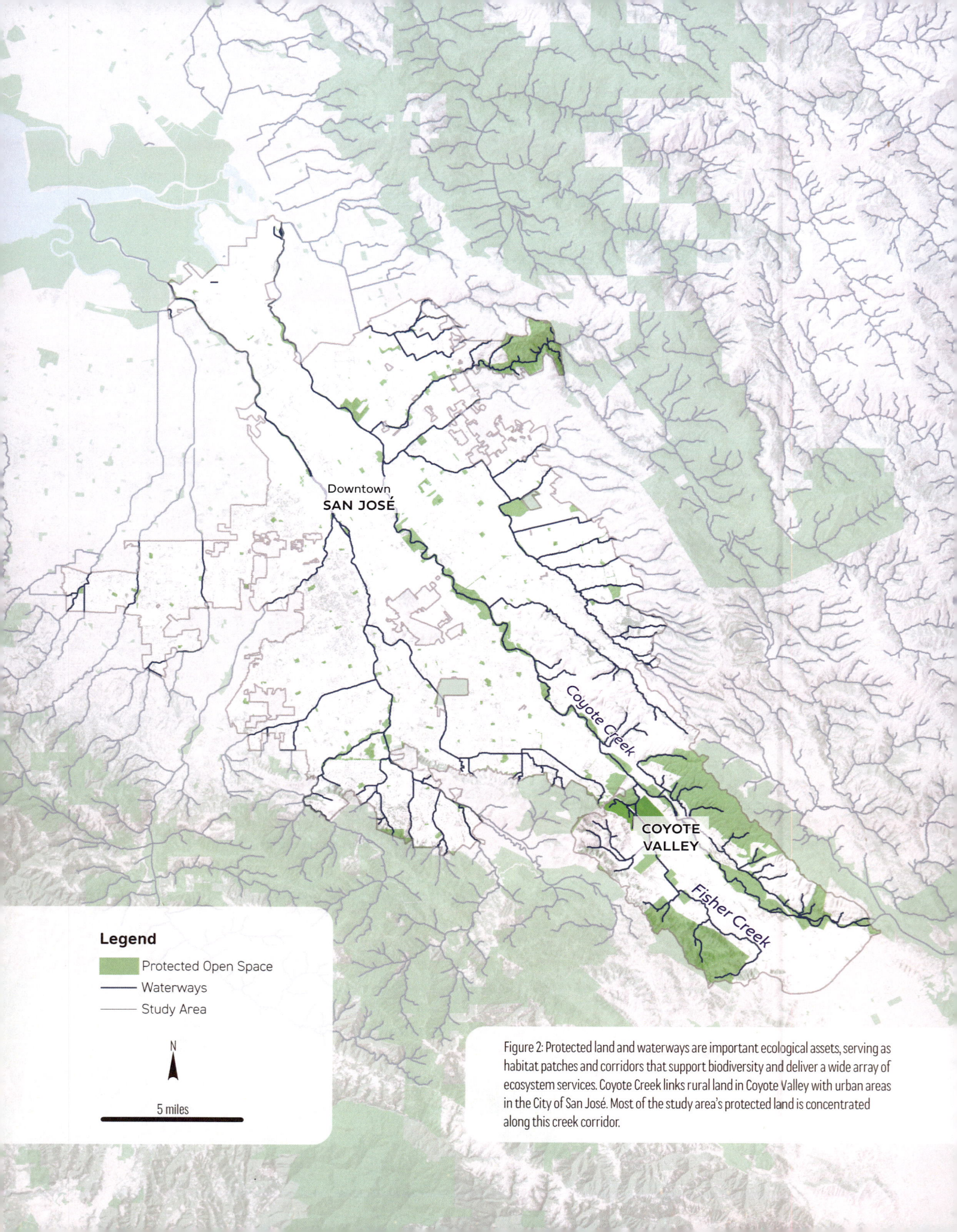

Legend

- Protected Open Space
- Waterways
- Study Area

N

5 miles

Downtown
SAN JOSÉ

COYOTE
VALLEY

Coyote Creek

Fisher Creek

Figure 2: Protected land and waterways are important ecological assets, serving as habitat patches and corridors that support biodiversity and deliver a wide array of ecosystem services. Coyote Creek links rural land in Coyote Valley with urban areas in the City of San José. Most of the study area's protected land is concentrated along this creek corridor.

▲ Downtown San José. Photograph by Eric Fredericks, courtesy of CC 2.0.

▲ Oblique view of part of San José. Photograph by Sergio Ruiz, courtesy of SPUR.

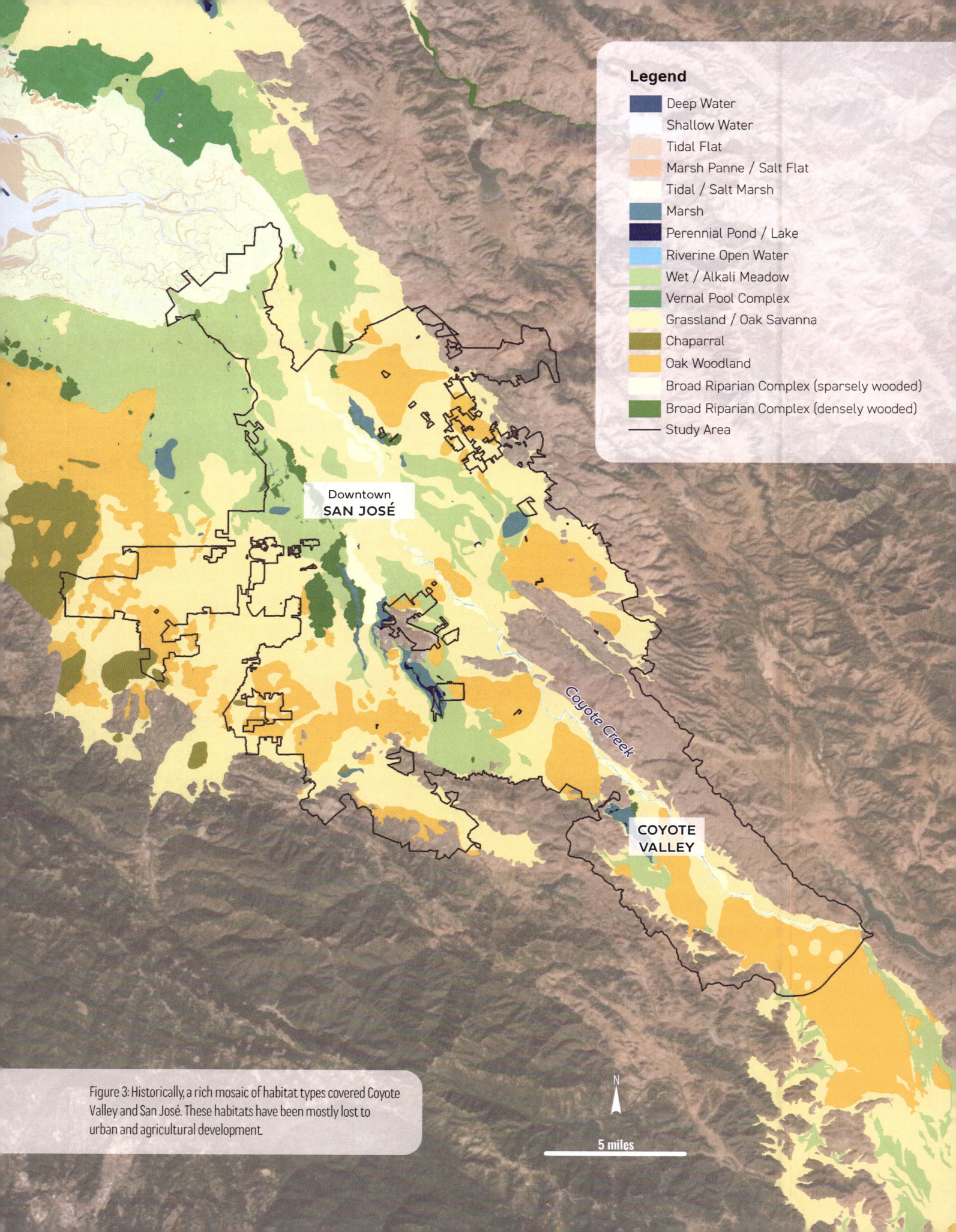

Legend

- Deep Water
- Shallow Water
- Tidal Flat
- Marsh Panne / Salt Flat
- Tidal / Salt Marsh
- Marsh
- Perennial Pond / Lake
- Riverine Open Water
- Wet / Alkali Meadow
- Vernal Pool Complex
- Grassland / Oak Savanna
- Chaparral
- Oak Woodland
- Broad Riparian Complex (sparsely wooded)
- Broad Riparian Complex (densely wooded)
- —— Study Area

Downtown
SAN JOSÉ

Coyote Creek

COYOTE VALLEY

Figure 3: Historically, a rich mosaic of habitat types covered Coyote Valley and San José. These habitats have been mostly lost to urban and agricultural development.

N

5 miles

Coyote Valley

South of San José, Coyote Valley narrowly separates the Santa Cruz Mountains and Diablo Range. The valley stretches from the southern extent of San José, through the unincorporated community of Coyote, to Morgan Hill. Coyote Creek, one of the principal waterways feeding the South Bay, runs the length of the valley. Historically, oak woodlands and savannas, sycamore alluvial woodlands, wet meadows, and dry grasslands blanketed much of the area. Laguna Seca, formerly one of the largest freshwater wetlands in the Bay Area, is located in the north end of Coyote Valley, south of Tulare Hill. This complex system of seasonal and perennial freshwater wetlands provides valuable habitat for a variety of wildlife (Grossinger et al. 2006).

Today, land in Coyote Valley is predominantly used for agriculture. With thousands of acres dedicated to orchards, ranches, row crops, and other types of farmland, the valley forms the largest remaining tract of prime farmland in Santa Clara County (FMMP 2018). Major crops include bell peppers, cherries, cabbage, and bok choy (SAGE 2012). Coyote Valley also contains rural housing developments and some industrial land uses, and serves as an important transportation corridor between the Bay Area and southern California. U.S. Highway 101 and the Union Pacific Railroad run parallel to Coyote Creek through much of the valley, as do two arterial roads: Monterey Highway and Santa Teresa Boulevard. Conversion of natural habitats to human land uses involved extensive modifications to the landscape, including digging drainage ditches, burning vegetation, and tilling, that opened much of the Valley to agriculture and reduced Laguna Seca to a seasonal wetland (Grossinger et al. 2006).

Despite the large-scale loss and alteration of natural habitats, Coyote Valley remains a critical habitat linkage for wildlife. Animals such as bobcats (*Lynx rufus*) and coyotes (*Canis latrans*) have been recorded traversing the valley to travel between protected areas in the Santa Cruz Mountains and Diablo Range (OSA 2017). Twelve species of rare, threatened, and endangered plants and animals reside in the valley, including the western burrowing owl (*Athene cunicularia hypugea*), Bay checkerspot butterfly (*Euphydryas editha bayensis*), and tricolored blackbird (*Agelaius tricolor*) (Thurlow 2019).

However, Coyote Valley's transportation corridors form major barriers to migrating wildlife, and pressure for urban development in the valley threatens its future habitat value (OSA 2017; OSA 2019). Actions to conserve existing habitat areas, such as the recent purchase of the North Coyote Valley property (Miller 2019), promise to protect resident wildlife while also providing benefits to downstream communities. Restoring and protecting the floodplains of Coyote Creek, for example, can slow the flow of stormwater through the creek— reducing flooding risk for communities along the creek in San José (OSA 2019). Increasing climate resilience for Coyote Valley and its downstream communities will also require actions on working lands, in industrial areas, and in residential communities.

NORTH COYOTE VALLEY

▲ Looking towards the Diablo Range with the North Coyote Valley Conservation Area in the foreground. Courtesy of OSA, photograph by R. Horii

In November 2019, 937 acres in Coyote Valley were permanently protected through an innovative public and private partnership among Peninsula Open Space Trust (POST), Santa Clara Valley Open Space Authority (OSA) and the City of San José. The $93.46 million acquisition deal was funded in part by Measure T, a $650 million infrastructure bond approved by San José voters in November 2018, which set aside $50 million for the purposes of conserving natural floodplains and sensitive groundwater areas in Coyote Valley. The recently conserved area presents a unique opportunity for restoring regionally rare habitat types, protecting critical wildlife linkages, and testing innovative climate resilience initiatives. Alongside key partners, Peninsula Open Space Trust and the City of San José, the Open Space Authority will lead a science and community-based planning process to establish a unique open space preserve and regional destination of statewide and national significance that preserves the environment, connects people to nature, and provides lasting climate resilience. The Plan will guide the future use and management of nearly 1,000 acres of open space within the North Coyote Valley Conservation Area, and will serve as a blueprint to implement OSA's Coyote Valley Landscape Linkage Report and achieve the floodplain preservation goals of San José's 2018 Measure T. One of the barriers to nature-based solutions is the need to better quantify their performance and benefits for comparison with traditional approaches. This community-based planning process is a significant opportunity to communicate the benefits of creating a public asset focused on interconnection, inclusion, and resilience, while designing with nature.

Study Area

Opportunities for nature-based solutions in the Bay Area vary depending on the local landscape and community priorities. This report uses SPUR's "place types" framework for landscape classification to identify different settings in which nature-based solutions can be applied. SPUR divided the nine-county Bay Area into a grid of half-mile squares and assigned each square a place type that describes its land use and physical form. This cluster analysis defined fourteen place types — such as Rural & Open Spaces, Cul-de-Sac Suburbs, Job Centers, etc. — based on housing density, job density, road intersection density, pavement permeability, and number of land uses. This framework is particularly useful for organizing landscape-scale strategies for greening that respond to the patchwork of site-specific conditions.

In an effort to test how nature-based solutions can be integrated into urban and rural areas in the Bay Area, we sampled place-types along a conceptual transect from downtown San José to Coyote Valley. Our study area includes urban place-types in the City of San José and rural place-types in Coyote Valley. We delineated Coyote Valley using the Metcalfe Canyon-Coyote Creek watershed (BAARI). Together San José and Coyote Valley capture nearly the full gamut of place-types found in the Bay Area, with the exception of four place-types found primarily in the denser cities of Oakland and San Francisco. In this report, we will focus on outlining nature-based solutions to address climate risks in six of the study area's place-types: urban neighborhoods, office parks, cul-de-sac suburbs, cultivated land, rural & open space, and parks & protected areas. These place-types vary widely in terms of housing, job density, and open space, yet they all face similar climate risks.

The following section outlines climate risks in San José and Coyote Valley, and chapter three explores in more detail the strategies that can increase climate resilience for people and benefit wildlife in these different place-types.

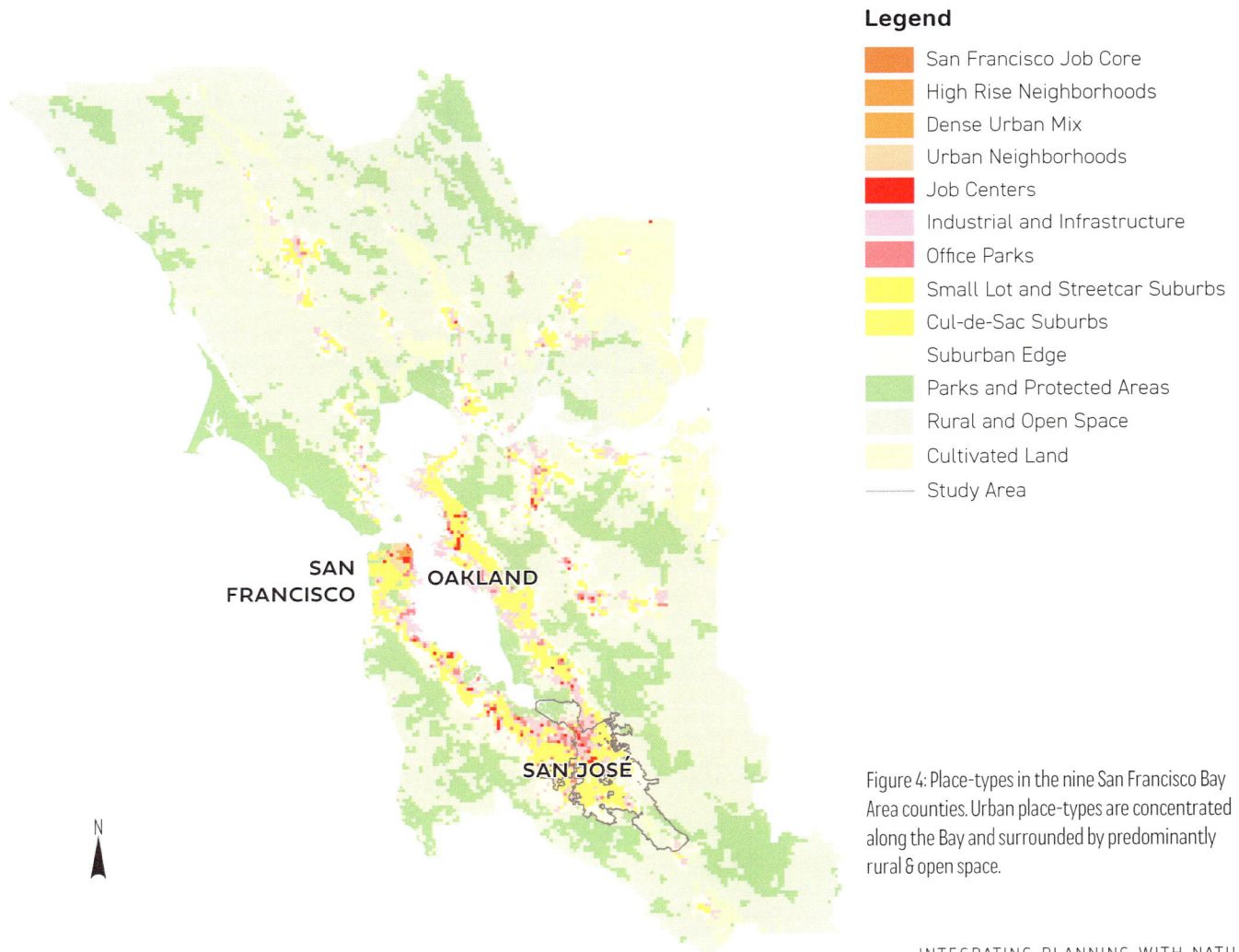

Legend

- San Francisco Job Core
- High Rise Neighborhoods
- Dense Urban Mix
- Urban Neighborhoods
- Job Centers
- Industrial and Infrastructure
- Office Parks
- Small Lot and Streetcar Suburbs
- Cul-de-Sac Suburbs
- Suburban Edge
- Parks and Protected Areas
- Rural and Open Space
- Cultivated Land
- Study Area

Figure 4: Place-types in the nine San Francisco Bay Area counties. Urban place-types are concentrated along the Bay and surrounded by predominantly rural & open space.

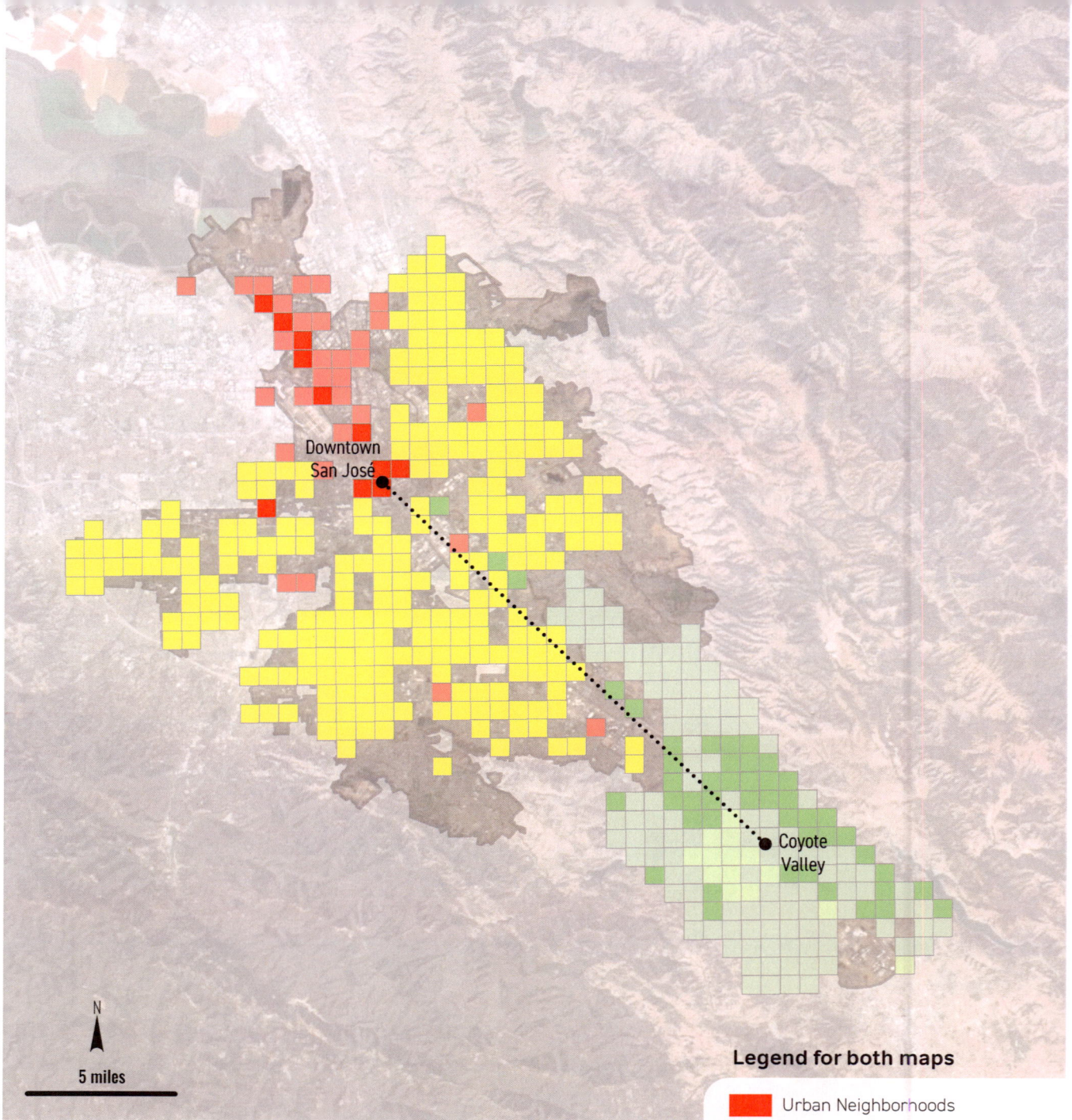

Figure 5: Focal place-types in study area. Urban neighborhoods, office parks, and agricultural land are concentrated along Coyote Creek and Guadalupe River. Cul-de-sac suburbs are the most commonly found place-type and are distributed throughout the City of San José. Protected land is largely located in the Diablo Range and links to rural & open space in the Santa Cruz Mountains through the valley floor.

Legend for both maps

🟥	Urban Neighborhoods
🟪	Industrial and Infrastructure
🟧	Office Parks
🟨	Cul-de-Sac Suburbs
⬜	Suburban Edge
🟩	Parks and Protected Areas
⬜	Rural and Open Space
🟩	Cultivated Land

Figure 6: Conceptual transect from downtown San José to Coyote Valley, moving from urban areas to rural land.

Hiking through Oak woodlands in Coyote Valley.
Photograph by D Mauk, courtesy of OSA.

Climate Risk

As global temperatures rise, temperatures in San José and Coyote Valley are likewise projected to increase, and rainfall is projected to become more sporadic (USGS 2014). These climatic shifts will result in several indirect climate risks, including increasingly intense droughts, floods, extreme heat events, and wildfires; declining air quality; and possible crop failure on agricultural lands (Fried, Torn, and Mills 2004; Jacob and Winner 2009; Mount et al. 2017; USGS 2014; Pathak et al. 2018). There is also the potential that gray infrastructure used to mitigate climate risk will further increase emissions, such as pumps used to deal with flooding or air conditioning during extreme heat events. These impacts are projected to disproportionately impact disadvantaged communities; achieving climate resilience will require addressing this inequity (Shonkoff et al. 2011; Schwarz et al. 2015; Cooley et al. 2016; Cushing et al. 2018). While sea-level rise will likely impact San José's bayshore, the shoreline falls outside the scope of this report. Additional resources are available to describe shoreline adaptation measures (e.g., Beagle et al. 2019).

In the absence of drastic changes to the City's use of fossil fuels, population growth will dramatically increase greenhouse gas emissions. The City of San José's 2016 review of its General Plan found that forecasted population and employment growth within San José will lead to community-wide increases in greenhouse gas emissions, primarily from the transportation sector (City of San José 2016). However, the City has taken bold steps forward by creating the San José Clean Energy Program, which provides the City's entire electrical grid with 86% renewable based energy, establishes Vehicle Miles Traveled thresholds for transportation impacts under CEQA, and sets a Vehicle Miles Traveled mitigation fee for future development projects. In partnership with the Santa Clara Valley Open Space Authority, the City is now developing a natural and working lands element for its climate action plan, Climate Smart San José, and is evaluating how avoiding sprawl development into greenfield areas like Coyote Valley could support denser transit-oriented infill and additional greenhouse gas avoidance. The County and OSA's Santa Clara Valley Agricultural Plan also examined how protecting the Santa Clara Valley's ranches and agricultural lands, like those in Coyote Valley, can avoid greenhouse gas emissions while providing opportunities to advance climate change goals by improving soil health and sequestering atmospheric carbon. The County has since launched the Agricultural Resilience Incentives grant program to fund stewardship practices on agricultural lands that increase beneficial ecosystem services, and is starting development of a Community Climate Action Plan to support local and regional collaboration to defend against climate risks (Girard et al. 2018). The following sections outline how San José and Coyote Valley are currently contributing to climate change, and provide additional details on each climate-related hazard.

Precipitation & Fluvial Flooding

While it is uncertain if total precipitation in San José and Coyote Valley will increase, decrease or stay the same, the seasonality of rainfall is likely to shift (USGS 2014). Wet seasons are likely to become shorter, more intense, and more variable while dry seasons become longer and hotter (USGS 2014; Berg and Hall 2015; Mount et al. 2017). Furthermore, climatic trends over the last century indicate that California is increasingly fluctuating between drought and extreme wet years (He and Gautam 2016). A range of climate scenarios predict that these fluctuations will become more severe into the future, large flood events will likely become more frequent and droughts will be more severe (Dettinger 2011; Mount et al. 2017).

Changes in precipitation will present a complex suite of challenges for residents of San José and Coyote Valley. More extreme rainfall events could result in more stormwater, sediment, nutrients, and trash being transported from the surrounding landscape into Coyote Creek and San José's other waterways. The catastrophic Coyote Creek flood of 2017, which prompted more than 14,000 people to evacuate and caused $100 million in damage, exemplifies the massive risk that increased stormwater can pose to San José's population and the importance of floodplains and open space areas as nature-based solutions for stormwater management (Romanow et al. 2018; Rogers 2019).

Interventions upstream in Coyote Valley, such as floodplain expansion, wetland restoration, and agriculture preservation, have the greatest water storage potential and benefit urban areas downstream. Reducing impervious cover and protecting open space in urban and rural areas can help reduce stormwater runoff. In addition, stormwater bioretention areas on streets, parks, and private parcels can capture flows, improve water quality, and reduce peak flooding.

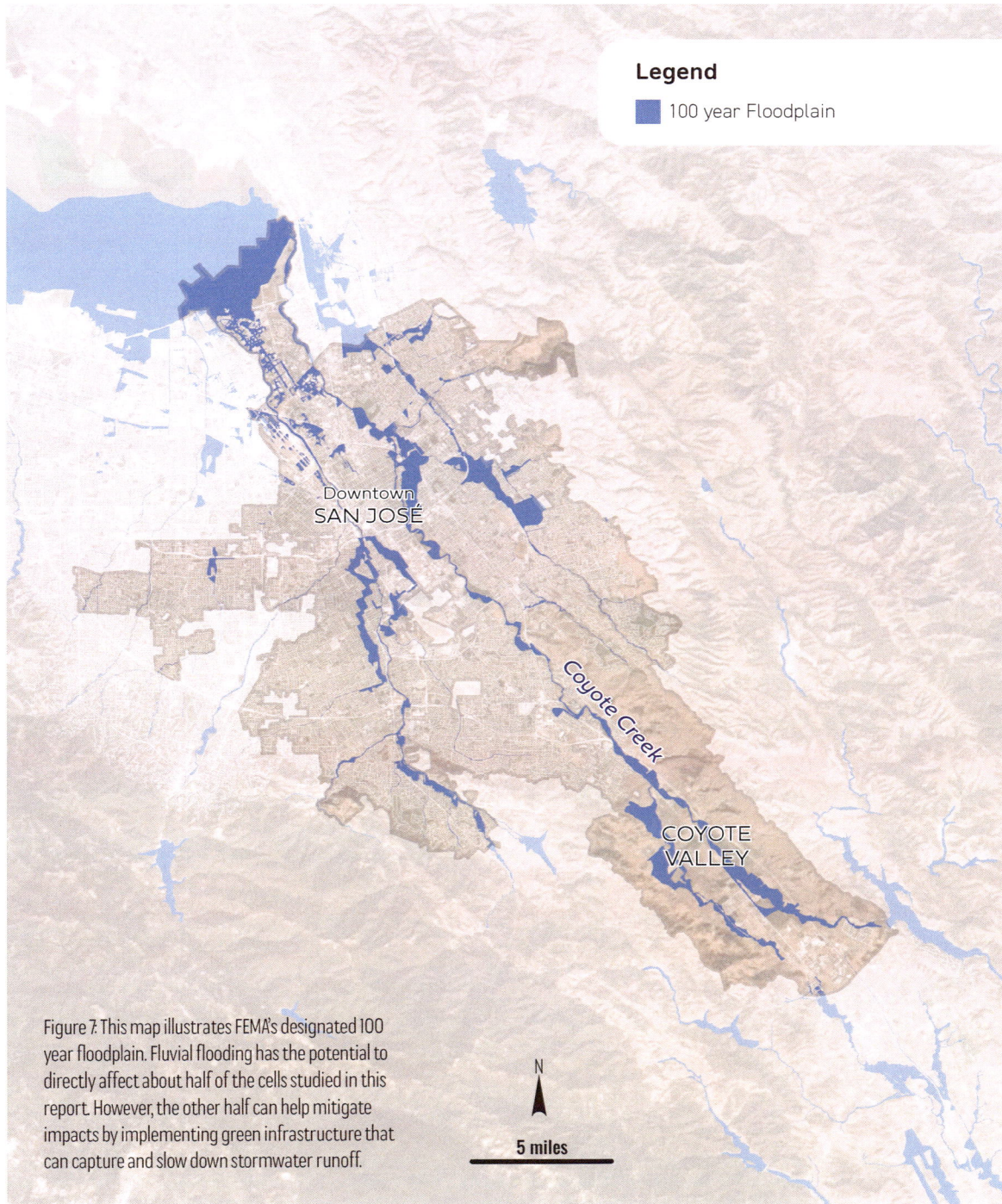

Legend
- 100 year Floodplain

Downtown SAN JOSÉ

Coyote Creek

COYOTE VALLEY

N

5 miles

Figure 7. This map illustrates FEMA's designated 100 year floodplain. Fluvial flooding has the potential to directly affect about half of the cells studied in this report. However, the other half can help mitigate impacts by implementing green infrastructure that can capture and slow down stormwater runoff.

Drought

Although future annual precipitation totals are uncertain, an increase in the length of the dry season in Santa Clara County is likely (USGS 2014; Berg and Hall 2015; Mount et al. 2017). Rising temperatures will increase the likelihood that low-precipitation years coincide with warmer summers, inducing drought (Diffenbaugh et al. 2015). The concentration of annual rainfall within fewer, more intense events could disproportionately increase overland runoff, and decrease the amount of rainwater that infiltrates and recharges groundwater supplies (Earman and Dettinger 2011).

The 2012-2016 drought demonstrated Bay Area cities' ability to address drought through water conservation measures, withdrawals from reservoirs and groundwater basins, and water purchases from agricultural water users and neighboring water systems (Lund et al. 2018). However, as climate change continues to add variability to California's water supply, cities such as San José will face increasing pressures on water-supply systems and must demonstrate continued adaptive management (Mount et al. 2017).

Groundwater is the sole water source for agricultural, domestic, municipal, and industrial water uses in Coyote Valley (Kassab et al. 2016). With longer dry seasons and higher summer temperatures, farmers in Coyote Valley will likely require additional water for irrigation and rely increasingly upon diminishing groundwater resources (Mount et al. 2017). While groundwater provided water users with a buffer against the 2012-2016 drought, more frequent and intense droughts may jeopardize the groundwater basin's future reliability as a water source (Diffenbaugh et al. 2015; Kassab et al. 2016). Future aquifer declines will not only impact groundwater-dependent ecosystems and creek flows, but also may impact the economies of Coyote Valley's rural communities, as occurred with communities in the Central Valley who faced higher levels of unemployment due to drought-related land fallowing in 2014-2015 (Howitt et al. 2015; Mount et al. 2017; Lund et al. 2018).

In addition to statewide and regional water policies and infrastructure improvements, several on-site nature-based improvements can help San José and Coyote Valley become more resilient to future droughts. Expanding riparian buffers, restoring floodplains and wetlands, installing stormwater retention and detention basins, and planting trees in both urban and rural place-types can slow the flow of stormwater and increase groundwater recharge rates (Sonneveld et al. 2018).

Rising Temperatures and Extreme Heat Events

As climate change increases average summertime temperatures in Santa Clara County, extreme heat events are likely to become more frequent. Current projections suggest their frequency will double by 2050 and triple by the end of the century in addition to lasting longer (USGS 2014; Cal-Adapt 2020). The California Energy Commission defines an extreme heat event as a day where the maximum daily temperature in a location is greater than 98% of the maximum temperatures that the location experienced between April and October 1961-1990 (Cal-Adapt 2020). This definition results in different temperature thresholds for different locations across the state to account for people's different levels of adaptation to heat (Vaidyanathan and Vaidyanathan 2013). The temperature thresholds for extreme heat events in downtown San José and Coyote Valley are 95.6°F (35.3°C) and 97.5°F (36.4°C), respectively (Cal-Adapt 2020).

Heat exposure can exacerbate a number of medical conditions, including cardiovascular risk, respiratory diseases, mental illnesses, stroke, and organ damage, and even lead to death (Hajat and Kosatky 2009). The heat wave that struck California in 2006 killed more than 600 people and resulted in 16,000 excess hospitalizations (Guirguis et al. 2014). In Santa Clara County Basu et al. (2008) found a 1.28% increase in mortality rate per 2°F (1°C) increase in air temperature. Heat events pose a serious risk to public health, especially to children, seniors, pregnant women, and other people with a diminished ability to thermoregulate (Kravchenko et al. 2013). Although the human body can acclimate to increases in temperature, the rapidity of climate change makes future acclimatisation uncertain. Heat waves that occur earlier in the year could also have more devastating effects.

Urban design can play an integral role in how often residents of a city experience extreme heat events. Urbanization and the loss of vegetation has altered urban microclimates. Impervious surfaces absorb and re-emit thermal energy from the sun more effectively than vegetation does in more natural settings (Oke 1982; Stone and Rodgers 2001). These altered thermal properties lead to cities being hotter than surrounding rural areas (Oke 1973). Additionally, reduced vegetation in urban areas prevents these areas from experiencing the cooling benefit of evapotranspiration (Oke 1982; Stone and Rodgers 2001). Tall buildings in cities can also block wind and provide additional surface area for absorbing solar heat (Oke 1982; Sakakibara 1996). Meanwhile, cars, air conditioning units, and industrial activity further release heat to the urban environment (Sailor 2011). As the climate changes, urban areas are therefore more

▲ Coyote Valley. Photograph by D. Neumann, courtesy of OSA.

susceptible to extreme heat events and their resultant health impacts (Corburn 2009). San José is particularly susceptible to the urban heat island effect, due to its high impervious cover and relatively low tree canopy cover.

Coyote Valley has little impervious cover compared to urbanized areas of San José and is thus less likely to suffer from the urban heat island effect, assuming open space is not developed. However, more frequent extreme heat events are nonetheless likely to impact the valley's workers and residents. California's farm workers complete arduous physical labor often in direct sunlight with limited opportunity for rehydration, making them particularly susceptible to heat-related illness. While state and federal regulations are in place to protect farm workers from heat-related illness, workers continue to show signs of heat stress (Moyce et al. 2017) and many workers report hydrating insufficiently (Stoecklin-Marois et al. 2013). These health impacts may become more commonplace without additional protections for farmworkers in the face of climate change.

Vegetation, particularly trees, can improve outdoor thermal comfort and safety and can reduce building energy consumption. Planted surfaces are more effective than high-albedo surfaces at reducing heat build-up and also present a number of co-benefits (Ong 2003). Tree canopy cover mitigates high temperatures by blocking incoming solar radiation and decreasing air temperature through evapotranspiration. Planting trees in urban and rural areas can help regulate local microclimates. New research shows that tree canopy cover equal to or greater than forty percent is necessary to significantly reduce temperatures at the block-scale in urban areas (Ziter et al. 2019). Trees have the potential to simultaneously reduce heat-related health impacts and air conditioning energy consumption, and should be recognized as critical infrastructure. In Sacramento, it has been estimated that tree shading reduces cooling energy demand by thirty percent (Gago et al. 2013). Trees can also be interspersed in cultivated land to provide areas of refuge for farm workers during hot days.

Trees shading active mobility corridors in San José State University. Photograph by Sergio Ruiz, courtesy of SPUR.

▲ Santa Clara County Firefighters.
Photograph by Daron L Wyatt (USFS), courtesy of CC 2.0.

Wildfire

As summer conditions become hotter and drier, wildfires are likely to become more frequent and more destructive across California (Fried et al. 2004; Westerling and Bryant 2008; Krawchuk and Moritz 2012). Future wildfires are likely to endanger human lives both directly for those living in their paths and indirectly for those exposed to their smoke plumes, which can extend for hundreds of miles (Tarnay 2018; Willingham 2018). Wildfire smoke exposure is associated with a variety of short- and long-term health impacts, including exacerbation of asthma and chronic obstructive pulmonary disease and increased all-cause mortality (Reid et al. 2016). Future wildfires will also endanger property and wildlife (Krawchuk and Moritz 2012) and impact water quality in California's waterways (Moody and Martin 2009; Coombs and Melack 2013).

While wildfire risk is relatively low within San José's urban core and Coyote Valley's highly modified farmlands, the areas will nonetheless face increased health risks from smoke produced in wildland fires (CalFire 2007; Reid et al. 2016). San José and other Northern California cities experienced some of the worst air quality in the world during the 2018 Camp Fire, when wildfire smoke reduced air quality to "unhealthy" and "very unhealthy" levels for a period of eleven days (Popovich et al. 2019). Wildfire risk is high on the Santa Cruz Mountains and Diablo Range that surround San José and Coyote Valley. The 2020 CZU and SCU Lightning Complex fires burned tens and hundreds of thousands of acres respectively of these mountain ranges. In these locations, fire poses a higher direct threat to human lives and the threat is likely to grow into the future (Westerling and Bryant 2008).

Human actions can largely determine the extent of wildfire damage as the climate changes (Mann et al. 2016). Land managers can reduce local fire risk by reducing fire ignitions, proactively managing land to prevent wildfires, prescribing grazing, and promoting land cover types more resistant to fire (e.g., shrublands and closed woodlands rather than invasive-dominated grasslands) (Keeley 2003; Bowman et al. 2011). Additionally, city planners can encourage higher density, low-impact development in more defensible urban centers, as opposed to expanding into the wildland-urban interface, where communities are more vulnerable (Davis 1990; Cohen 2000). Protecting open space in rural areas is not only crucial in curbing sprawl and reducing the number of people and communities at risk, but also provides connections for wildlife populations to escape wildfires in the adjacent mountain ranges.

Air Pollution

Climate change may also affect the concentrations of air pollutants other than smoke in urban and rural settings. The air pollutants of primary concern for human health are ozone and particulate matter (Jacob and Winner 2009), both of which are associated with increased emergency department visits, hospitalizations, and deaths due to respiratory and cardiovascular diseases (Kheirbek et al. 2013).

Concentrations of surface ozone generally increase with increased air temperature (Jacob and Winner 2009). Ozone typically forms when volatile organic compounds (VOCs) react with nitrogen oxides (NO_x). Automobiles and industrial activity are generally the primary sources of NO_x, while vegetation is often a major source of VOCs (Fitzky et al. 2019). In urban and high-traffic areas, NO_x levels are typically high and VOC concentrations limit the production of ozone (Calfapietra et al. 2013). Under a changing climate, the Bay Area is projected to experience marked increases in ozone due to increased biogenic VOC production at higher temperatures (Weaver et al. 2009). Reducing vehicular traffic and NO_x emissions can, however, mitigate this effect (Fitzky et al. 2019).

The effect of climate change on particulate matter concentrations is uncertain. The frequency of precipitation and the amount of atmospheric mixing are major drivers of local particulate matter concentrations, and it remains unclear how they will change under a warming climate (Jacob and Winner 2009). The principal sources for particulate matter in Santa Clara County are construction, farming operations, domestic fuel combustion, and passenger vehicles (Fanai et al. 2014). Reducing passenger vehicle traffic and increasing the use of renewable energy can decrease particulate matter pollution in the county's urban areas (Harlan and Ruddell 2011). Green walls, trees, and hedges planted along streets can reduce street-level particulate matter and NO_x (Pugh et al. 2012). Meanwhile, adopting farming practices that reduce airborne dust production (e.g., conservation tillage and manual harvesting) can improve air quality on agricultural lands (Clausnitzer and Singer 1996; Arslan and Aybek 2012).

▲ Dust from harvest. Photograph by Frank Shepherd, courtesy of CC 2.0.

Mulching can help retain soil moisture. Photograph courtesy of USDA.

Reduced Agricultural Productivity

Agriculture in Coyote Valley and elsewhere in California may become increasingly volatile as variable rainfall, drought conditions, and warmer temperatures impact crop yields. Increasing prevalence of agricultural pests and diseases and more volatile costs of agricultural inputs (e.g., fertilizers and energy) may contribute to further variability in the agricultural sector. Coyote Valley's various crops are likely to respond differently to climatic changes given their different temperature and irrigation requirements (Pathak et al. 2018). Of the crops currently grown in Coyote Valley, cherries may be the most vulnerable to warming (SAGE 2012; Pathak et al. 2018). Higher temperatures and extreme heat may render growing certain crops, such as stone fruit, untenable in Coyote Valley and elsewhere in California (DeJong 2005). Cherries and other stone fruits require cold winter temperatures to break their dormancy, and warmer winter temperatures lead to a smaller and lower-quality fruit yield (Pathak et al. 2018). Coyote Valley's rangelands are likewise vulnerable to shifting temperature and precipitation regimes. Shorter growing seasons for grasslands and intermittent dry years may result in inadequate forage for livestock (Chaplin-Kramer and George 2013). Conversely, temperature changes may allow crops grown elsewhere in the state to grow well in Coyote Valley, and more climate-resilient crops may remain profitable (Jackson et al. 2011).

Adopting climate-smart agriculture practices, such as mulching and compost application, can help retain soil moisture and buffer the impacts of variable precipitation and rising temperatures. These measures also reduce irrigation demand (potentially benefiting neighboring urban areas that depend on the same water supply) and improve soil and plant health. In addition, adding hedgerows and riparian buffers that support beneficial insects can become part of an integrated pest management strategy that reduces the need for expensive chemical inputs and combats the increasing prevalence of agricultural pests.

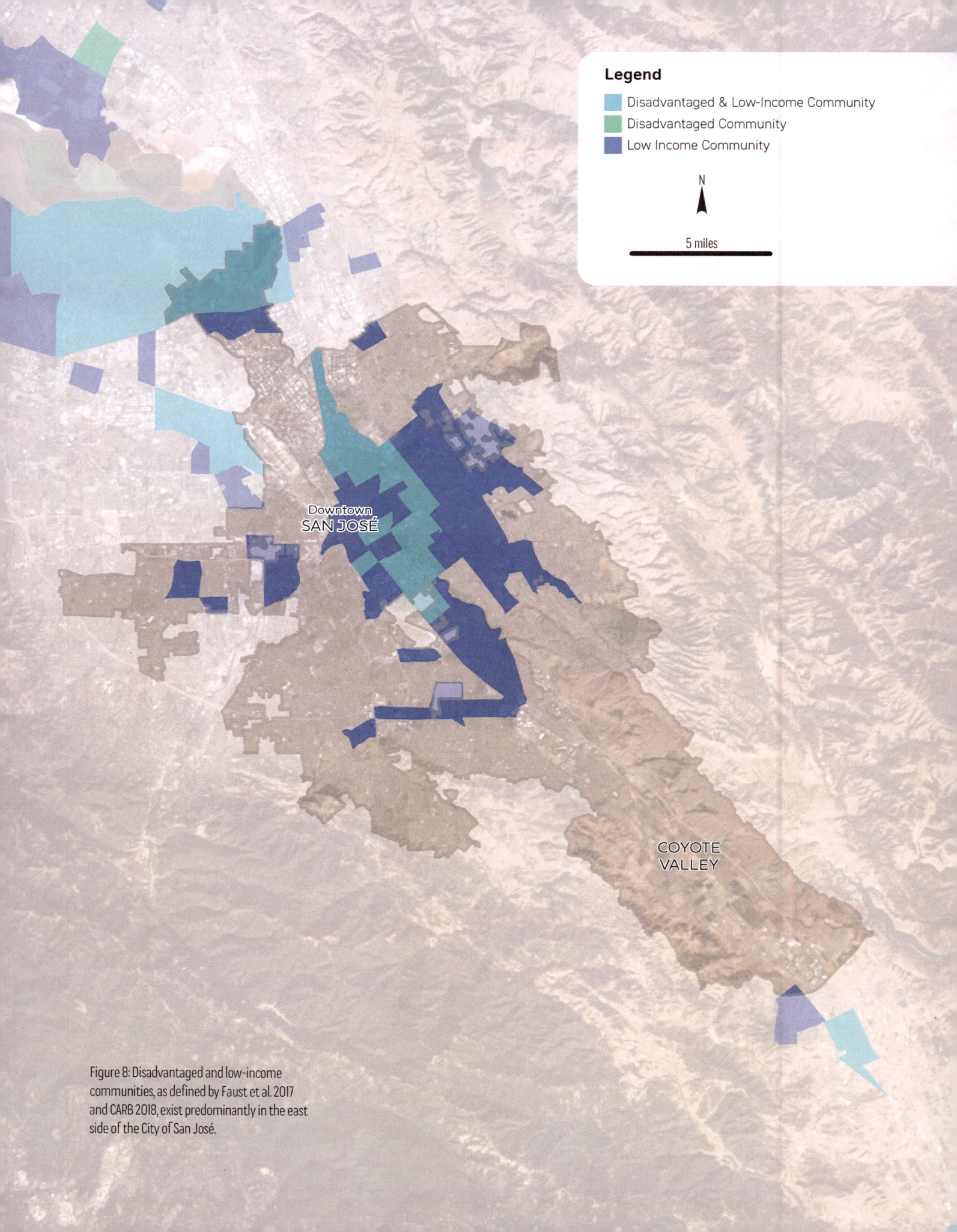

Legend

- Disadvantaged & Low-Income Community
- Disadvantaged Community
- Low Income Community

N

5 miles

Downtown
SAN JOSÉ

COYOTE
VALLEY

Figure 8: Disadvantaged and low-income
communities, as defined by Faust et al. 2017
and CARB 2018, exist predominantly in the east
side of the City of San José.

Environmental Injustice

The impacts of climate change are likely to disproportionately affect disadvantaged communities in San José. Low-income communities are already located in areas where air pollution (Faust et al. 2017) and fluvial flooding risk tend to be higher (FEMA 2009). Park access and tree canopy cover in Santa Clara County is lower in low-income neighborhoods, a disparity that is also associated with race. White households, in contrast with Latino and Asian households, are located closer to parks and in areas of higher tree canopy cover regardless of income (see page 38). Given the benefits conferred by parks and tree canopy cover, low-income and, in particular, communities of color are more vulnerable to climate change. This is consistent with research that shows that underserved neighborhoods are more susceptible to the urban heat island effect (Stone and Rodgers 2001; Schwarz et al. 2015). Existing health disparities in Santa Clara County (e.g., higher asthma and obesity rates in low-income neighborhoods) may further exacerbate the inequity of climate change's health effects (Santa Clara County 2016).

The economic effects of climate change may also impact disadvantaged communities more than more affluent neighborhoods. Water rates, for example, tend to rise during droughts and may become prohibitively expensive for low-income households (Cooley et al. 2016). Low-income households are also less likely to be able to financially withstand property losses from flooding or fires, and increases in food prices may impact their food security (Shonkoff et al. 2011).

One way in which the state identifies disadvantaged communities is by considering health and socioeconomic indicators alongside pollution levels (Faust et al. 2017). It separately designates low-income communities as census tracts and households at or below 80% of the statewide median income, or at or below a threshold set by the California Department of Housing and Community Development (CARB 2018). Figure 8 displays disadvantaged and low-income communities in and around San José.

California allocates a percentage of its cap-and-trade revenue for projects that benefit environmental health in disadvantaged or low-income communities. California currently stipulates that at least 25% of funds from its cap-and-trade program must support projects within and benefitting disadvantaged communities (as defined by Faust et al. 2017), and an additional 10% must support low-income households or communities (CalEPA 2018). While this policy aims to address disparities in climate risk, studies have demonstrated that cap-and-trade has simultaneously led to increased air pollution in disadvantaged communities (Cushing et al. 2018). Critics also argue that CalEnviroScreen (CalEnviroScreen 3.0) fails to identify all disadvantaged communities in San José (Dueñas 2016, and see further discussion below).

It is important to note that the state uses the above-mentioned varying definitions of disadvantaged communities depending on the context in which they are applied. The definition based on CalEnviroScreen includes a household income factor, but brings in other factors including environmental burdens. The other commonly used definition is based on 80% of statewide median household income (MHI). Based on these metrics, Santa Clara County appears relatively affluent, with a median household income 63% higher than the statewide value in 2019 (US Census Bureau 2018). But given the high cost of living - especially the high cost of housing - in the study area, neither statewide definition adequately represents the economic disadvantages which residents in the region experience. Further work is therefore necessary to achieve climate justice in the city.

As efforts continue at the state level to establish more meaningful regional definitions of disadvantaged communities, a local resource is available to help understand the relevant disadvantages of South Bay communities, especially as they regard access to nature. The Open Space Authority's *Understanding our Community* report (Olson et al. 2016) defines "Deep Engagement Communities" that identify elevated economic, linguistic, transportation, and environmental burdens and barriers.

DISPARITIES IN TREE CANOPY COVER AND PARK ACCESS

Comparing demographic data for Santa Clara County with data on urban greenery illuminates multiple disparities in environmental benefits along lines of race and income. The most populous races and ethnicities in the county are White, Hispanic and Latinx, and Asian or Pacific Islander (US Census Bureau 2017). While higher-income households across these three groups tend to be located within areas of higher canopy, the average canopy surrounding white households of any income bracket exceeds the averages of all but the wealthiest bracket of Asian and Pacific Islander or Hispanic and Latinx households. Tree canopy provides shade that can help reduce urban heat (Gago et al. 2013), and is also associated with reductions in violence and aggression (Kondo et al. 2017), stress relief (Beil and Hanes 2013), lower risk of cardiovascular disease (Donovan et al. 2013), increase longevity (Takano et al. 2002), and other health benefits. The disparity in Santa Clara County's tree canopy distribution signifies that lower-income and non-white households are less likely to receive these cooling and health benefits (Salmond et al. 2016). Similarly, across these three races and ethnicities, higher income households tend to be located in areas with more nearby parkland. Within each income bracket, white households tend to have the greatest acreage of greenspace within one mile. This disparity diminishes the accessibility of outdoor recreation for lower-income and non-white households, and likely contributes to local health disparities (Abercrombie et al. 2008; Santa Clara County 2011). Appendix 3 contains a detailed description of the methods used to obtain these results.

▼ East San José. Courtesy of Google Earth.

Neighborhood in Atherton, Santa Clara County at same scale. ▲
Courtesy of Google Earth.

Canopy Cover by Income and Race

Figure 9: Tree canopy cover in Santa Clara County by income and race. Census tracts with predominantly white households had higher tree canopy regardless of income level.

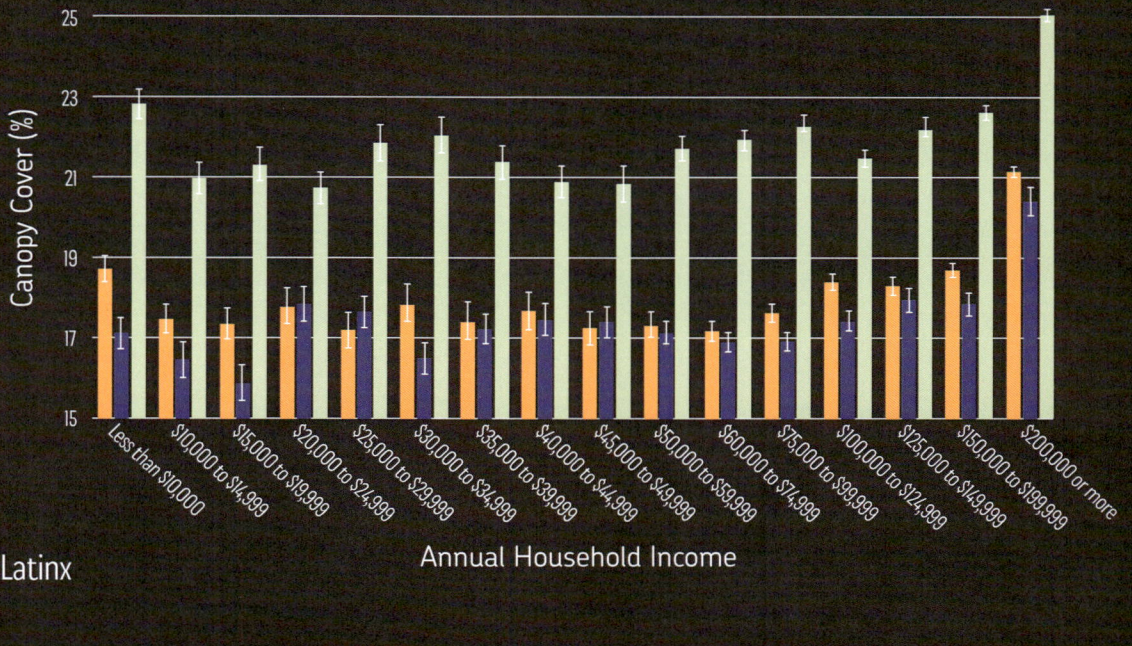

- Asian
- Hispanic or Latinx
- White

Park Area within 1 Mile by Income and Race

Figure 10: Park area within a mile buffer of households in Santa Clara County by income and race. Census tracts with predominantly white households had access to more park space within a mile radius regardless of income level.

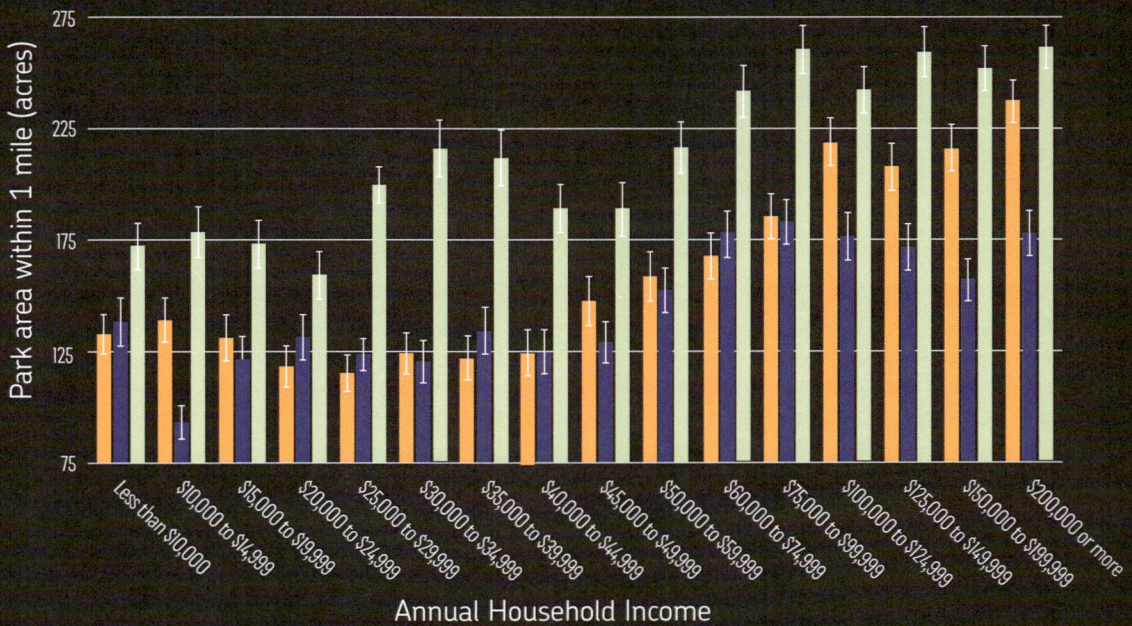

3
IDENTIFYING OPPORTUNITIES AND PATHWAYS

The Bay Area is already facing a severe housing shortage, and population is expected to increase by as much as 4 million residents and 2.1 million jobs in the next 50 years[1]. There is the risk that cities will grow at the expense of nature, further debilitating an already weakened ecological system and increasing the region's vulnerability to climate change.

───────────

1   The Center for Continuing Study of the California Economy provided SPUR with population and job projections as detailed in their report, "High and Low Projections of Jobs and Population for the Bay Area to 2070 -Projection Framework, Specific Assumptions and Results". The report included a "high growth projection" and a "low growth projection," based on national projections for jobs and population as well as assumptions about immigration, growth in various economic sectors, and the share of the population and job growth that the Bay Area will attract. SPUR has chosen to base its analysis on the high growth projection in order to determine the number of housing units needed to meet population growth. The housing analysis was conducted by the Concord Group for SPUR.

Left, top: 4th and San Fernando Street in San José, California. Photograph by Sergio Ruiz, courtesy of SPUR.
Left, bottom: Tract house in Suisun City. Photograph by ClatieK, courtesy of CC 2.0.

Land in the nine Bay Area counties is predominantly rural and low density. About 84% of the land is classified as one of the three rural place-types identified by SPUR in the *Bay Area Regional Strategy* (Parks & Protected Areas, Rural & Open Space, or Cultivated Lands). Only 26% of the Bay Area's rural land is protected, with the remainder subject to varying degrees of development pressure. The greatest threat to open space in the Bay Area is the expansion of subdivision development. In urban areas, about 75% of land is zoned for single-family residential neighborhoods, which together represent 69% of the region's housing stock. Compact, mixed-use, walkable neighborhoods serviced by mass transit make up only 1% of urban land. However, 22% of urbanized areas are low-density commercial areas, which present the greatest potential for sustainable redevelopment. A coordinated regional strategy for resilience will need careful consideration of how all place-types can together contribute to climate change adaptation and mitigation.

SPUR's *Bay Area Regional Strategy* presents a vision for how the region can grow over the next half century by increasing density in existing urban areas. The report examines what housing and job growth would look like in six place-types: Dense Urban Mix, Urban Neighborhoods, Office Parks, Industrial & Infrastructure, Small-Lot & Streetcar Suburbs, and Cul-de-Sac Suburbs through scenario modeling. These future higher density scenarios or 'Model Places,' are conceptual illustrations of densification based on idealized place-type cells. They demonstrate that there is room for growth and that regional goals can be achieved within the existing urban footprint by the year 2070. The 'Model Places' served as a template for our study of nature-based solutions in increasingly dense urban areas. 'Model Places' do not exist for rural place-types, since all projected growth was located in urban areas, so we selected existing cells in Coyote Valley as a canvas. The following housing and job projections for each place-type is based on SPUR's Regional Strategy.

We selected three place-types (urban neighborhoods, office parks, and cul-de-sac suburbs) to typify high, medium, and low-density areas in San José and three place-types (parks & protected areas, rural & open space, and cultivated lands) to characterize Coyote Valley's landscape typologies. The following place-type scenarios are the product of a co-discovery and co-development process with a diverse group of expert stakeholders. We hosted two hands-on workshops, one on future greener scenarios of urban place-types and another for rural place-types. Participants were asked to suggest big moves for the integration of nature-based solutions within these areas to enhance climate resilience. They included key staff from SFEI, SPUR, and OSA as well as representatives from:

- City of San José Public Works, Parks, Planning, Sustainability, Transportation and Environmental Services departments;
- Santa Clara Valley Habitat Agency;
- Santa Clara County Office of Sustainability, Parks Department, and Department of Planning and Development;
- California Department of Conservation and Office of Planning and Research;
- Google;
- Valley Water;
- Design and consulting firms including AECOM, CMG, Sitelab, HT Harvey & Associates, Studio Tsquare, HMH, and Environmental Science Associates; and
- Nonprofit organizations including The Nature Conservancy, Committee for Green Foothills, and Guadalupe River Park Conservancy.

Overall, there was great alignment among participants on the importance of investing in natural infrastructure in urban and rural place-types, increasing density near transportation nodes, protecting open space, restoring native biodiversity, conserving agricultural land, and collaborating across disciplines. The following sections describe the bold, collective visions for incorporating nature-based solutions in each place type that were explored during the two workshops, planning and policy approaches for implementing these strategies, and quantification of some of the benefits provided by implementing these strategies.

Bird's eye view of Palo Alto. Photograph by Sergio Ruiz, courtesy of SPUR.

Coyote Valley. Photograph by B Adams, courtesy of OSA.

Urban Neighborhood Scenario

This place-type marks the highest density end of the urban-to-rural gradient in our study area. Urban neighborhoods offer a mixture of jobs, housing, and supporting services, which facilitates strategic planning for mass transit routes and therefore sustainable densification. There are eleven of these 0.5 mile by 0.5 mile place-type cells (1,750 acres) in San José, including the downtown area. They are characterized by high impervious cover (ranging between 56-81%), varying building cover (11-40%), and very low tree canopy cover (6-18%). Across the Bay Area, urban neighborhoods are forecast to accommodate 157,000 new residents, or 5,425 per cell.

Greening interventions in higher density areas have the potential to benefit the greatest number of people. Tightly-spaced buildings make for more walkable neighborhoods but also increase the importance of placing nature-based solutions within the public right-of-way. By using trees and bioretention features along the existing road network, San José can create shaded active mobility corridors that simultaneously increase ecological connectivity, intercept stormwater runoff, and filter air pollutants. While competition for ground-level land uses may render construction of new urban greenspaces difficult, they nonetheless provide outsized ecological and health benefits. New parks should be designed to accommodate the community's social, ecological, and functional needs as much as possible.

Key strategies for integrating nature-based solutions

Parks: Create a large park near civic facilities and prioritize new publicly accessible greenspaces in park-poor neighborhoods. Improve park access and ecological connectivity of greenspaces by distributing them roughly within a half mile of each other. Increase canopy cover in parks to at least 45% cover to create park cool islands that can also reduce temperatures in neighboring blocks. Plant local native species to provide greater biodiversity support and reduce water consumption.

Urban forest: Incorporate trees along all streets to improve outdoor thermal comfort, increase habitat connectivity, and capture air pollutants. Increase overall canopy cover in private and public land to 40% or more for cooling benefits. Plant local native species to provide greater biodiversity support and reduce water consumption. Increase available soil volume for trees to promote tree health, accelerate canopy growth, and increase stormwater detention. Preserve existing large trees that provide ecological value to wildlife.

Bioretention systems: Build centralized stormwater detention and filtration ponds in low-lying areas with capture cisterns. Add bioretention areas along streets and at intersections to store, treat, and reduce road runoff. Integrate trees to increase soil water storage. Increase permeability as much as possible by replacing driveways and parking lots with pervious materials.

Green roofs and walls: Prioritize green roofs and walls on buildings facing greenspaces and in areas with low softscape cover. Green roofs located on lower levels will provide greater ecological connectivity with ground-level habitat. Intensive green roofs, i.e., those with deeper soil profiles which can accommodate shrubs and trees, will have greater benefits.

2020

0.5 miles

2070

0.5 miles

Legend

- Existing greenspace*
- New greenspace*
- Existing tree
- New tree
- Building
- New Building
- Street

*Publicly accessible

◀ Based on draft urban design courtesy of AECOM for SPUR, December, 2019. Final designs available in the *Model Places* report (SPUR and AECOM, 2020).

2070

Shuttle/Train

Shuttle/Train

0.5 miles

Legend

- ▨ Existing greenspace*
- ▧ New greenspace*
- ● Existing tree
- ● New tree
- ▢ Building
- ▨ New Building
- ╱ Street

*Publicly accessible

1. Community park in former intersection
2. Transit Oriented Development
3. Centralized stormwater basin in new public greenspace
4. Shaded bike and pedestrian street with Emergency Vehicle Access
5. New parks created through strategic acquisition

6. Trees on all streets
7. New job and transit center with affordable housing units
8. Vegetative highway buffer
9. Publicly accessible greenspace in townhome complex with mid-block passage

Benefit quantification

- Greenspace area tripled while doubling the number of residents and jobs housed in this cell.
- A combination of a central civic green, pocket parks and publicly accessible greenspaces in new developments make up twelve additional acres of greenspace.

Housing Units

2020 4,084 existing units

2070 Increase of 5,425 units (total of 9,509 units)

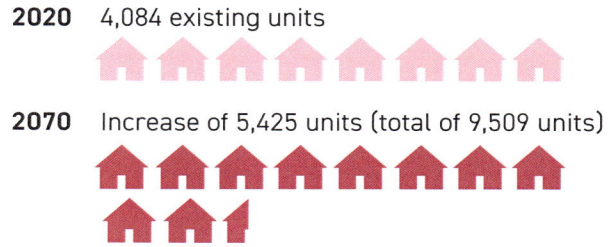

Jobs

2020 2,881 existing jobs

2070 Increase of 3,746 jobs (total of 6,627 jobs)

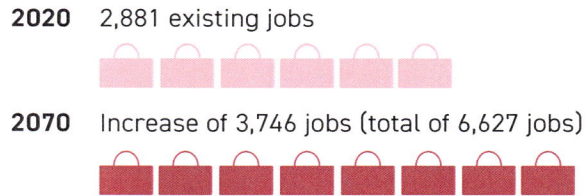

Greenspace

2020 4.5 existing acres

2070 Increase of 12 acres (total of 16.5 acres)

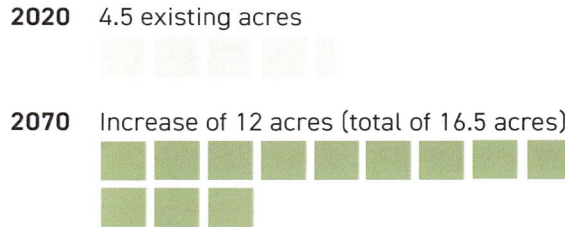

Trees

2020 2,745 existing trees

2070 Increase of 1,170 trees (total of 3,915 trees)

Public Greenspace in San José's urban neighborhoods ranges between 0-26 acres per cell (out of 160 acres total).

The Model Place currently has 4.5 acres. New developments with taller buildings, road diets, and public-private partnerships make room for 7.8 additional acres of publicly accessible greenspace, totaling 12.3 acres.

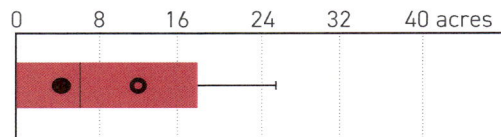

Legend for both Public Greenspace and Tree Canopy Cover diagrams

- ⊢▬⊣ All SJ cells (n=11)
- ● 2020 Model Place
- ○ 2070 Model Place

Tree Canopy Cover varies in San José's urban neighborhoods between 6-18%.

The Model Place currently has 15% canopy cover. The addition of 1,170 trees and growth of existing trees increases canopy cover to 47%, conferring significant cooling benefits.

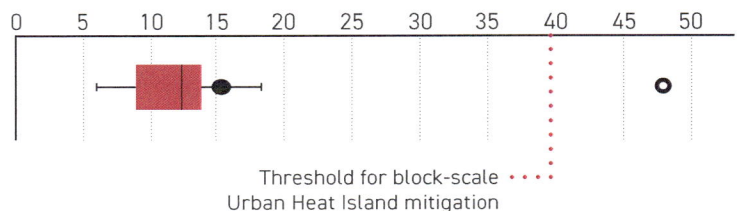

Threshold for block-scale Urban Heat Island mitigation

Urban Neighborhood Scenario *(continued)*

Changes to the urban forest can provide valuable ecosystem services. In order to preserve existing mature trees and reduce street tree mortality rates, cities will need to invest in maintenance and design new planting areas to ensure tree health. The choice of tree species and vegetative structure is critical (see Appendix 2). The future scenario assumes that new plantings and street tree replacements will draw from local native species that can best support native wildlife and are well adapted to San José's climate. Ecosystem service comparison of urban forest below using iTree (see Appendix 1).

Carbon Sequestration (tons/year)

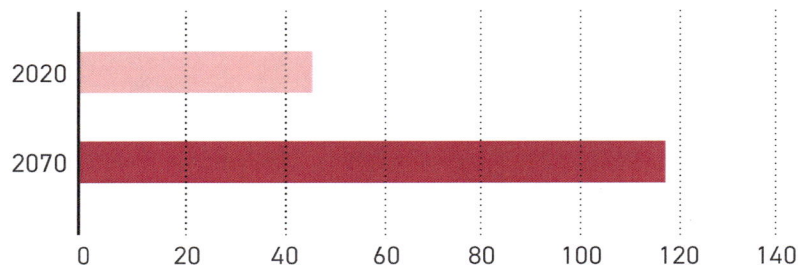

Total Carbon Storage (tons)

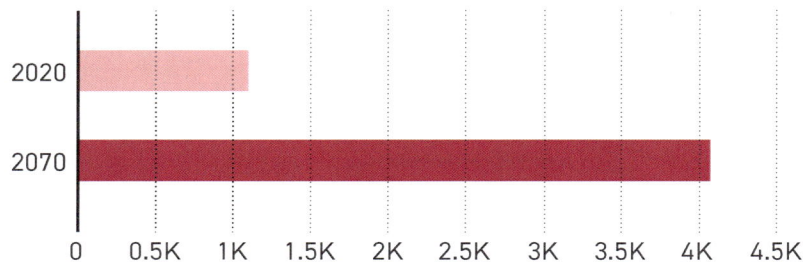

Total Air Pollution Removal (lbs/year)

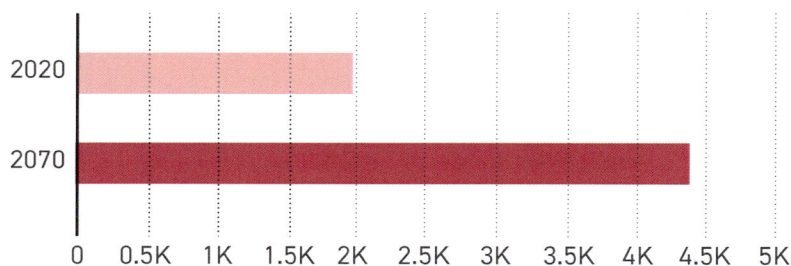

Avoided Runoff (cubic feet/year)

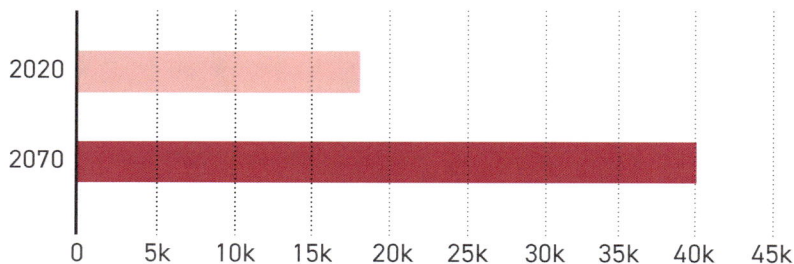

Planning and policy approaches for implementation

Community based planning: Promote community-driven planning with robust engagement processes. Consult stakeholder groups early in the process to set goals and priorities, as well as gather local spatial knowledge. Explore opportunities for community-based nonprofits to identify and organize neighborhood actions.

Environmental justice: Introduce policies to prevent displacement of underserved communities, so that existing communities can benefit from greening interventions. Develop housing and job equity plans in conjunction with major greening efforts. Acquire undeveloped lots in park-poor areas to build new greenspaces.

Outreach and education: Site new stormwater features in highly trafficked areas and schools to serve as educational opportunities.

Transportation: Articulate street network hierarchy to identify which streets can be pedestrianized and accommodate larger landscape areas. Add non-automotive corridors to connect job centers, commercial areas, and retail spaces with surrounding neighborhoods. Eliminate parking minimums to promote walkability, free up space for bioretention features, and reduce the area of surface parking lots.

Greenspace regulation: Leverage new development to fund publicly accessible greenspace creation and improvement by offering incentives (e.g., floor-area-ratio bonuses) and commercial park fee requirements. Introduce green area ratio (weighted proportion of greenspace in a lot) regulations for development permit applications. Develop a city-wide plan to deliver publicly accessible greenspaces within a ten minute walk of all residents.

Tree removal and replacement: Set high tree replacement ratios (i.e., the number of trees that must be planted for each existing tree removed) to increase overall canopy cover. Increase requirements for tree basin size that promote tree health and longer lifespans. Protect existing large and healthy trees that provide ecological and environmental benefits. Promote the use of native and site-appropriate tree species to better support local biodiversity and reduce water consumption.

Design guidelines: Establish park, streetscape, and stormwater design guidelines that prioritize local native species, mimic native habitat structure, buffer wetlands, and promote wildlife-friendly management practices.

Urban nature planning: Develop a master plan that can guide new development by identifying cool islands and hot spots, major sources of air pollution, areas prone to flooding, ecological corridors, habitat zones, and priority areas for park creation.

Utilities: Create policies to move utilities underground for any redevelopment, including high-end development projects and existing residential neighborhoods.

Office Park Scenario

The single-use zoning and car-oriented approach to planning office parks in Silicon Valley, and in many other regions of the United States, resulted in the development of big box buildings surrounded by extensive surface parking lots. The prioritization of vehicular transportation makes office parks not only unfriendly to pedestrian circulation, but also exacerbates the generation of heat islands, air pollution, and stormwater runoff. Investments in more accessible and less carbon intensive transportation options is critical to the redevelopment of office parks and remains one of the greatest challenges to short-term improvements. Nevertheless, the underutilization of space and consolidation of private ownership now makes office parks one of the place-types with the greatest potential for large-scale redevelopment and greenspace creation.

Office parks in the nine Bay Area counties are forecast to accommodate 110,000 new housing units and 315,00 jobs. The thirty office-park cells (4,800 acres) in San José have greater impervious cover (ranging between 52-87%), less building cover (3-32%), and more variable tree canopy cover (5-29%) than urban neighborhoods. They are usually located along highways, arterial streets, and stream corridors.

Key strategies for integrating nature-based solutions

Parks: Replace impervious surfaces with multi-functional publicly accessible greenspaces that offer greater recreational opportunities, cool the neighborhood, and provide wildlife habitat. Consolidate buildings around a transit hub to free up more greenspace. Arrange greenspaces to create a greenbelt around development in relation to existing features and the riparian corridor.

Greenway: Build a densely forested corridor that combines stormwater capture and active mobility trails along the border of the existing office park and residential area.

Riparian corridors: Restore the riparian corridor and channel to make room for seasonal flooding, capture pollutants, and improve ecological connectivity. Limit wind turbines near wetlands and stream corridors to reduce bird collisions.

Urban forest: Increase overall tree canopy cover at a district scale from 8 to 45% to mitigate the urban heat island effect. Create active mobility corridors with double rows of trees that encourage people to walk and bike throughout the area. Plant local native species to provide greater biodiversity support and reduce water consumption. Increase available soil volume for trees to promote tree health, accelerate canopy growth, and increase stormwater detention. Preserve existing large trees that provide ecological value to wildlife.

Bioretention systems: Add large water detention features in parks for seasonal flows that can also serve as amphitheaters and playing fields during the dry season. Integrate features in streets and plazas to capture localized runoff.

Green roofs: Integrate photovoltaic panels and vegetation on building roofs. Prioritize green roofs and walls on buildings facing greenspaces and in areas with low softscape cover. Green roofs located on lower levels will provide greater ecological connectivity with ground-level habitat. Intensive green roofs, i.e., those with deeper soil profiles which can accommodate shrubs and trees, will have greater benefits.

Surface parking lots can be repurposed as large connected greenspaces.

2020

0.5 miles

2070

0.5 miles

Legend

Existing greenspace*

New greenspace*

Water

Existing tree

New tree

Building

New Building

Street

*Publicly accessible

◀ Based on draft urban design courtesy of AECOM for SPUR, December, 2019. Final designs available in the *Model Places* report (SPUR and AECOM, 2020).

Office Park Scenario *(continued)*

2070

Bike/Ped

Bike/Ped

Shuttle/Train

Shuttle/Train

Bike/Ped

Bike/Ped

0.5 miles

Legend

- ▢ Existing greenspace*
- ▢ New greenspace*
- ▢ Water
- ● Existing tree
- ● New tree
- ▢ Building
- ▢ New Building
- — Street

*Publicly accessible

1 Pedestrian and bicycle greenway doubles as ecological corridor

2 Shaded pedestrian streets connect job center to residential neighborhood

3 Transit hub located along a grand boulevard complete with bioretention features

4 A combination of vegetable gardens, flexible amphitheaters, detention basins, and contemplative natural areas encircle the development

5 Development rights along the riparian corridor are transfered to areas away from waterways to make room for floodplain restoration

6 Existing private greenspaces are redesigned with native plants and higher tree canopy cover

7 Parking garages are located at the edges and replace extensive surface parking lots

Benefit quantification

- Buildings are consolidated near the transit hub and away from the riparian corridor making room for 37 new acres of greenspace and 2,000 additional trees.
- These improvements provide 15 acres of greenspace per 1,000 residents or five times greater than the state of California's target for park provision.

Housing Units

2020 476 existing units

2070 Increase of 632 units (total of 1,108 units)

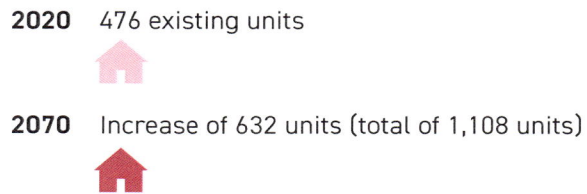

Jobs

2020 3,297 existing jobs

2070 Increase of 1,808 jobs (total of 5,105 jobs)

Greenspace

2020 0 existing units

2070 Increase of 37 acres (total of 37 acres)

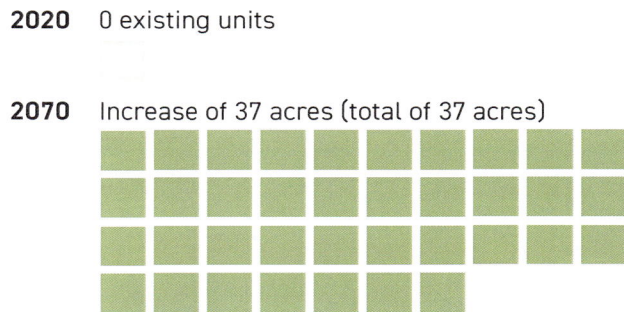

Trees

2020 1,441 existing trees

2070 Increase of 2,000 trees (total of 3,441 trees)

Public Greenspace in San José's
Public Greenspace in San José's office parks ranges between 0-30 acres per cell
(out of 160 acres total).

The Model Place currently has 0 acres. Taller buildings and district parking structures make room for 37 additional acres of publicly accessible greenspace.

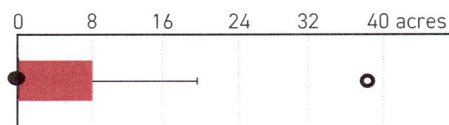

Legend for both Public Greenspace and Tree Canopy Cover diagrams
├─■─┤ All SJ cells (n=11)
● 2020 Model Place
○ 2070 Model Place

Tree Canopy Cover varies between 5-29%.

The Model Place currently has 8% canopy cover. The addition of 2,200 trees and growth of existing trees increases canopy cover to 46%.

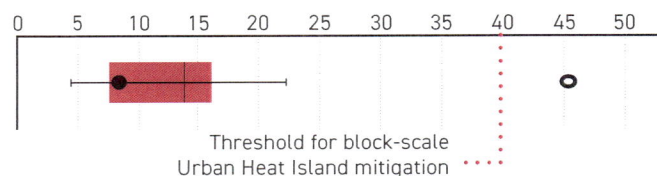

Threshold for block-scale
Urban Heat Island mitigation

Office Park Scenario *(continued)*

Changes to the urban forest can provide valuable ecosystem services. In order to preserve existing mature trees and reduce street tree mortality rates, cities will need to invest in maintenance and design new planting areas to ensure tree health. The choice of tree species and vegetative structure is critical (see Appendix 2). The future scenario assumes that new plantings and street tree replacements will draw from local native species that can best support native wildlife and are well adapted to San José's climate. Ecosystem service comparison of urban forest below using iTree (see Appendix 1).

Carbon Sequestration (tons/year)

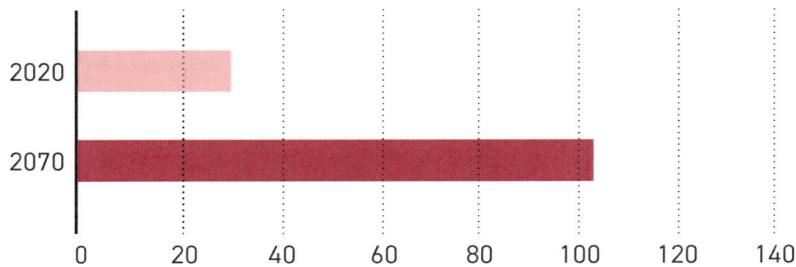

Year	
2020	~30
2070	~103

(x-axis: 0, 20, 40, 60, 80, 100, 120, 140)

Total Carbon Storage (tons)

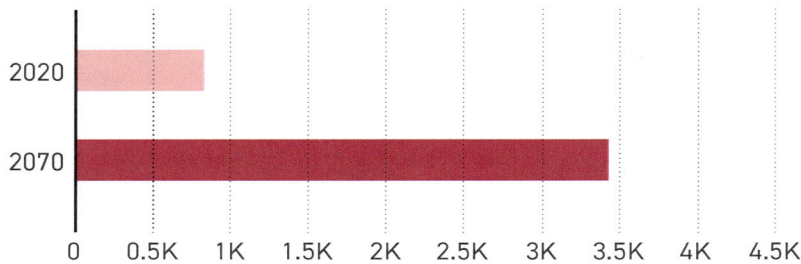

Year	
2020	~0.8K
2070	~3.4K

(x-axis: 0, 0.5K, 1K, 1.5K, 2K, 2.5K, 3K, 3.5K, 4K, 4.5K)

Total Air Pollution Removal (lbs/year)

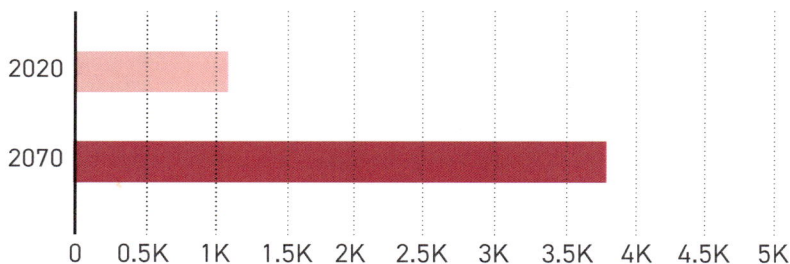

Year	
2020	~1.1K
2070	~3.8K

(x-axis: 0, 0.5K, 1K, 1.5K, 2K, 2.5K, 3K, 3.5K, 4K, 4.5K, 5K)

Avoided Runoff (cubic feet/year)

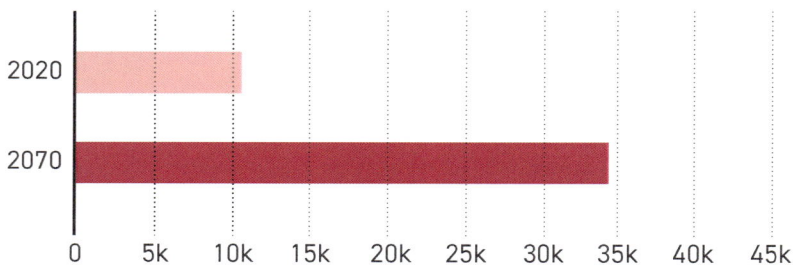

Year	
2020	~10k
2070	~34k

(x-axis: 0, 5k, 10k, 15k, 20k, 25k, 30k, 35k, 40k, 45k)

Planning and policy approaches for implementation

Transportation: Drastically reduce, or eliminate, surface parking lots and consolidate parking to district structures. Limit vehicular circulation at the center of the office district to facilitate pedestrian streets. Reduce the number and width of road lanes to free up ground-level space for bioretention systems.

Incentives: Create incentives for greenspace creation, such as floor-area-ratio bonuses, tax credits, and green rebates.

Greenspace regulation: Leverage new development to fund publicly accessible greenspace creation and improvement by offering incentives (e.g., floor-area-ratio bonuses and tax credits) and commercial park fee requirements. Introduce green area ratio (weighted proportion of greenspace in a lot) regulations for development permit applications.

Exchange markets: Create transferable development rights programs to increase building setbacks along riparian corridors and concentrate development around transit nodes and other designated growth areas.

Development fees: Add regulation and fees on new developments to help pay for green stormwater improvements. Introduce commercial fee requirements to help fund park creation.

Tree removal and replacement: Set high tree replacement ratios (i.e., the number of trees that must be planted for each existing tree removed) to increase overall canopy cover. Increase requirements for tree basin size that promote tree health and longer lifespans. Protect existing large and healthy trees that provide ecological and environmental benefits. Promote the use of native and site-appropriate tree species to better support local biodiversity and reduce water consumption.

Zoning: Encourage policies that modify existing zoning to facilitate a mixture of land uses and reduce commute times.

Overlay zones: Create habitat protection overlay zones along sensitive habitat, such as wetlands and stream corridors, to limit building encroachment, light pollution, and other environmental stressors.

Design guidelines: Establish park, streetscape, and stormwater design guidelines that prioritize local native species, mimic native habitat structure, buffer wetlands, and promote wildlife-friendly management practices. Set bird-safe and wildlife-friendly architecture requirements (e.g., minimize nighttime lighting, apply exterior treatment on windows, select anti-glare materials) on new developments and building retrofits.

> ## "Double the density with half the footprint, double the open space."
>
> -Kevin Conger (CMG)

Cul-de-Sac Suburb Scenario

Following World War II, federal policies facilitated the mass production of cul-de-sac suburbs to increase homeownership in the white middle class and stimulate the economy (Dreier et al. 2004). These policies not only resulted in large-scale disinvestment in cities but also spatial segregation by race and class (Frumkin 2002). The characteristic single-family home and private yard has become intimately tied to ideals of home ownership in the United States and changes to this place-type will require a shift in cultural attitudes and extensive outreach campaigns. The availability of private yards is part of the allure of suburban living, but these have been traditionally designed as manicured non-native gardens that often lack biodiversity and ecosystem-service benefits. In this place-type, existing greenspaces and road networks present the greatest opportunity for the integration of nature-based solutions and can be coupled with innovative transportation strategies.

Cul-de-sac suburbs are the most common place-type in San José, totalling 252 cells (40,320 acres). The characteristic single-family home and yard typology that make up cul-de-sac suburbs leads to relatively lower impervious cover (ranging between 39-79%), reduced building cover (11-31%), and variable tree canopy cover (4-34%). SPUR's Regional Strategy estimates cul-de-sac suburbs are forecast to accommodate 655,000 housing units and 160,000 jobs in the Bay Area. In order to increase density, cul-de-sac suburbs will need shuttles and other shared transportation alternatives that connect these neighborhoods to regional transit corridors.

Key strategies for integrating nature-based solutions

Riparian corridor: Move structures away from hydrological features, expand floodplain and riparian corridor to reduce property damage due to flooding, capture pollutants, and improve wildlife connections.

Greenways: Build forested corridors, with trails when possible, behind homes to improve pedestrian circulation and ecological connectivity.

Urban forest: Build grand boulevards with multi-modal transportation shaded by rows of large native trees to encourage active mobility, mitigate urban heat islands, capture air pollution, and sequester carbon.

Bioretention systems: Add detention and retention ponds in larger open areas and small bioretention features along streets to capture, store, and treat stormwater.

Parks: Enhance the ecological value and reduce the water consumption of existing greenspaces by implementing native habitat restoration strategies, such as re-oaking. Build 'seed' parks in neighborhoods to pilot initiatives and garner community support.

Front and back yard improvements: Plant locally native and drought-tolerant vegetation to support local wildlife and reduce water consumption. Increase permeable cover to promote water infiltration and reduce stormwater runoff. Plant trees to expand the urban forest and associated ecosystem services.

Bioretention systems: Add large water detention features in parks for seasonal flows that can also serve as amphitheaters and playing fields during the dry season.

Green roofs: Integrate photovoltaic panels and vegetation on building roofs.

2020

0.5 miles

2070

0.5 miles

Legend

- ⬜ Existing greenspace*
- 🟩 New greenspace*
- 🟦 Water
- 🟢 Existing tree
- ⬤ New tree
- ⬜ Building
- 🟨 New Building
- ━ Street

*Publicly accessible

◀ Based on draft urban design courtesy of AECOM for SPUR, December, 2019. Final designs available in the *Model Places* report (SPUR and AECOM, 2020).

2070

Shuttle

Bike/ Ped

Shuttle

Ped

Ped

Ped

Bike/ Ped

0.5 miles

Legend

Existing greenspace*

New greenspace*

Water

● Existing tree

● New tree

□ Building

New Building

Street

*Publicly accessible

1 New public greenspaces developed as part of new townhome and apartment complex

2 Strategic buyouts of buildings along riparian corridor give way to restoration and flood-risk reduction

3 Former parking lot becomes a community garden

4 Homes at the end of cul-de-sacs are bought back to create new neighborhood parks and pedestrian shortcuts

5 Tree canopy corridors and active mobility trails improve ecological connectivity and provide shortcuts

6 Streets are lined with double rows of trees and bioretention features

Benefit quantification

- Moderate population growth, fifteen new acres of greenspace, and 1,430 additional trees are accommodated through a combination of riparian restoration, park creation, and street redesign.
- Greenspace area per capita is more than twice the state of California's target.

Housing Units

2020 2,187 existing units

2070 Increase of 563 units (total of 2,750 units)

Jobs

2020 754 existing jobs

2070 Increase of 137 jobs (total of 891 jobs)

Greenspace

2020 6 existing acres

2070 Increase of 15 acres (total of 21 acres)

Trees

2020 2,382 existing trees

2070 Increase of 1,430 trees (total of 3,812 trees)

Public Greenspace in San José's

cul-de-sac suburbs ranges between 0–53 acres per cell (out of 160 acres total).

The Model Place currently has 6 acres. 15 acres of publicly accessible greenspace are created by strategically buying out homes in the floodplain and through public-private partnerships.

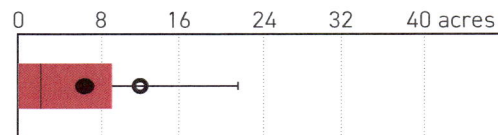

0   8   16   24   32   40 acres

Legend for both Public Greenspace and Tree Canopy Cover diagrams
— All SJ cells (n=11)
● 2020 Model Place
○ 2070 Model Place

Tree Canopy Cover varies

between 4–34%.

The Model Place currently has 18% canopy cover. The addition of 1,500 trees and growth of existing trees increases canopy cover to 43%.

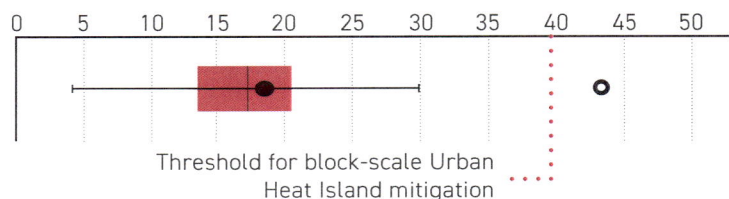

0   5   10   15   20   25   30   35   40   45   50

Threshold for block-scale Urban Heat Island mitigation

Cul-de-Sac Suburb Scenario *(continued)*

Changes to the urban forest can provide valuable ecosystem services. In order to preserve existing mature trees and reduce street tree mortality rates, cities will need to invest in maintenance and design new planting areas to ensure tree health. The choice of tree species and vegetative structure is critical (see Appendix 2). The future scenario assumes that new plantings and street tree replacements will draw from local native species that can best support native wildlife and are well adapted to San José's climate. Ecosystem service comparison of urban forest below using iTree (see Appendix 1).

Carbon Sequestration (tons/year)

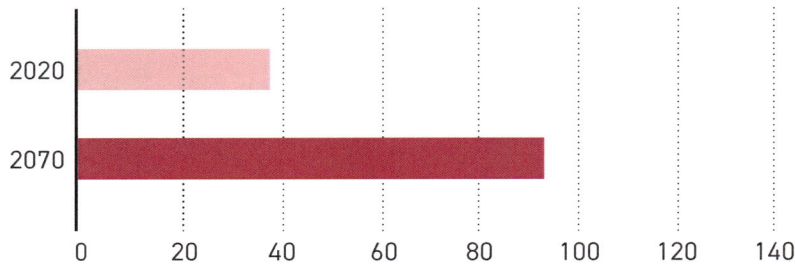

Year	
2020	~38
2070	~93

(x-axis: 0, 20, 40, 60, 80, 100, 120, 140)

Total Carbon Storage (tons)

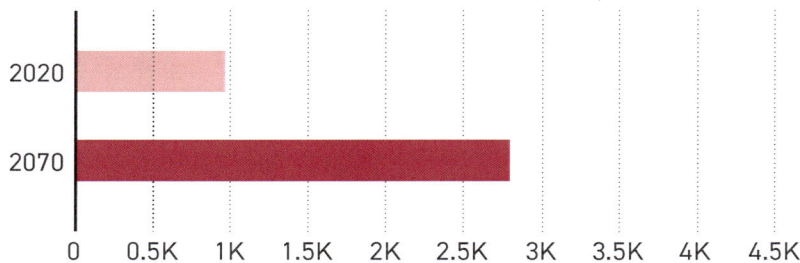

Year	
2020	~1K
2070	~2.8K

(x-axis: 0, 0.5K, 1K, 1.5K, 2K, 2.5K, 3K, 3.5K, 4K, 4.5K)

Total Air Pollution Removal (lbs/year)

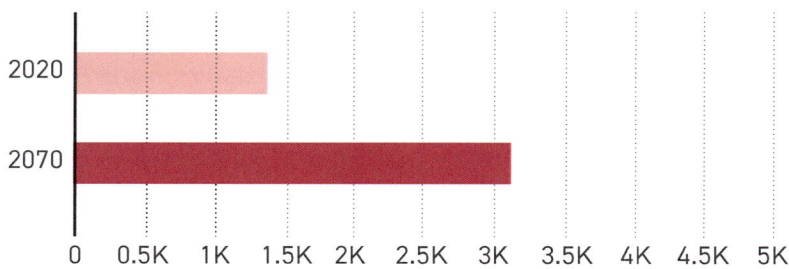

Year	
2020	~1.4K
2070	~3.1K

(x-axis: 0, 0.5K, 1K, 1.5K, 2K, 2.5K, 3K, 3.5K, 4K, 4.5K, 5K)

Avoided Runoff (cubic feet/year)

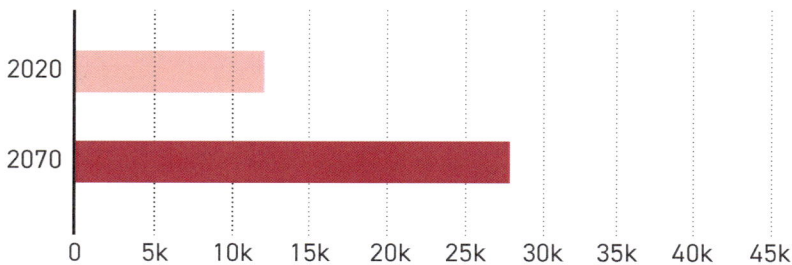

Year	
2020	~12k
2070	~28k

(x-axis: 0, 5k, 10k, 15k, 20k, 25k, 30k, 35k, 40k, 45k)

Planning and policy approaches for implementation

Zoning: Encourage policies that modify existing zoning to facilitate a mixture of land uses (e.g., multi-family units, retail, commercial, and office space) and reduce commute times. Promote infill development within existing building footprints, when possible and appropriate, to avoid greenspace loss. Lots near transportation nodes can be converted to apartment buildings to diversify housing stock and secondary housing units can be developed as part of an existing home (e.g., elder cottage housing built over a garage).

Density targets: Work with local stakeholders to set minimum and maximum density targets that are sensitive to community priorities, enhance the feasibility and profitability of mass transit, and reduce development pressures in the urban periphery.

Buyout programs: Develop programs to buy back private land at market rates. Strategic buyouts of parcels located in existing and projected floodplains can help prevent damages and restore ecosystem function. Cities and community land trusts can also acquire undeveloped lots in park-poor neighborhoods to build publicly accessible greenspaces that provide recreational opportunities and serve as nodes of ecosystem service delivery. In suburbs, parcels at the end of cul-de-sacs can be repurposed as gateways to forested trail networks that improve pedestrian circulation and facilitate wildlife movement. Federal programs (e.g., Flood Mitigation Assistance Program and HMA grants) and state and local bonds (e.g., Measure T) can help fund the strategic acquisition of land to meet resilience and habitat goals.

Outreach and education: Create a robust community outreach strategy that involves residents and Home Owner Associations (HOAs) early in the planning process. Provide evidence of the benefits of increasing density to raise support for changes in planning controls (e.g., reduction in HOA fees, the opportunity to age in place and raise funding for community assets). Develop long-term public education programs around nature-based solutions and ecosystem services to catalyze their implementation. Prioritize locating nature-based solutions in highly visible areas and near schools to serve as educational demonstration projects.

Greenspace regulation: Leverage new development to fund publicly accessible greenspace creation and improvement by offering incentives (e.g., floor-area-ratio bonuses and tax credits) and commercial park fee requirements. Introduce green area ratio (weighted proportion of greenspace in a lot) regulations for development permit applications. Create transferable development rights programs to move buildings away from riparian corridors and concentrate development around transit nodes and other designated growth centers.

Homeowner incentives: Offer tax incentives for homeowners to increase permeable cover, plant and protect trees, select native species, and maintain greenspace in their private property. Organize native plant giveaway events that make it easier for people to access site-appropriate plant palettes.

Cultivated Land Scenario

Urban growth has often come at the expense of cultivated lands. In the past thirty years, Santa Clara County has lost 21,171 acres of its farm and rangeland to development (Girard et al. 2018). The remaining 28,391 acres are at risk of being converted to office parks and single-family homes. These productive landscapes are vital local food sources and important rural job centers. Cultivated lands also provide ecosystem benefits, such as foraging habitat, wildlife linkages, groundwater recharge, flood risk reduction, and more.

However, intensive agricultural practices, including monoculture cultivation, use of chemical inputs, and frequent land disturbance, diminish the ecosystem benefits gleaned from cultivated lands. Climate-smart agricultural practices, such as cover copping, compost application, reduced tilling, hedgerows, and mulching can enhance the ecological value of our working lands without compromising local food production. Restoring habitat even in small unused agricultural areas can yield significant benefits to humans and wildlife. Coordinated action is needed to keep agriculture viable near urban areas and to enhance its ecological value.

There are 18 cultivated land cells (2,880 acres) in Coyote Valley concentrated on the valley floor. 13 of these cells, or 72%, are located along a creek and could implement measures to boost riparian function. Overall, land cover in these cells is predominantly dedicated to the cultivation of grain, row crops, hay, or pasture (77%), followed by industrial development (5%) and rural residential uses (5%). Tree canopy cover is very low and ranges between 0.1 to 6.6%.

Key strategies for integrating nature-based solutions

Riparian corridors and buffers: Restore historic creek alignment and riparian woodland. Plant riparian buffers in fields adjacent to creeks to capture nutrients and pollutants.

Hedgerows: Establish rows of shrubs, grasses, or other perennial vegetation between fields and greenhouses to provide nesting, forage, and shelter for pollinators, mammals, birds, reptiles, and amphibians. Hedgerows can also prevent wind-driven erosion, capture air pollutants, and sequester and store carbon. Coordinate hedgerow locations between different parcels to create continuous wildlife corridors through the valley floor.

Mulching: Apply mulch in agricultural fields and orchards to increase soil and below ground carbon sequestration, reduce erosion, protect soil from compaction, improve moisture retention, reduce water use, and suppress weeds.

Bioretention systems: Add infiltration and detention basins beside greenhouses to treat stormwater locally.

Management: Adopt wildlife-friendly farming practices (e.g., adding raptor boxes, reducing use of pesticides and fertilizers) that also improve water quality and capture particulate matter.

Crop switching: Transition to more profitable crops (e.g., heirloom wheat and hemp) that stand a better chance of adapting to projected climatic conditions.

Cover crops: Plant grasses, legumes, forbs, and other groundcover crops between rows or underneath orchards to reduce erosion from wind and water, increase biodiversity, regulate soil moisture, and improve the soil's ability to store and sequester carbon.

Re-oaking: Plant trees, especially native oaks, along all roads to create shaded ecological corridors. Intersperse trees in farmland to create a place of refuge for farm workers on hot days.

Compost application: Apply compost to croplands to improve soil health and increase soil microbial organisms and plant biomass, and therefore increase carbon sequestration and storage.

Reduce tilling: Limit or eliminate tilling to improve air quality, as well as preserve topsoil, maintain or improve soil health and quality, increase plant-available moisture, and reduce fossil fuel-derived powered machinery.

2020

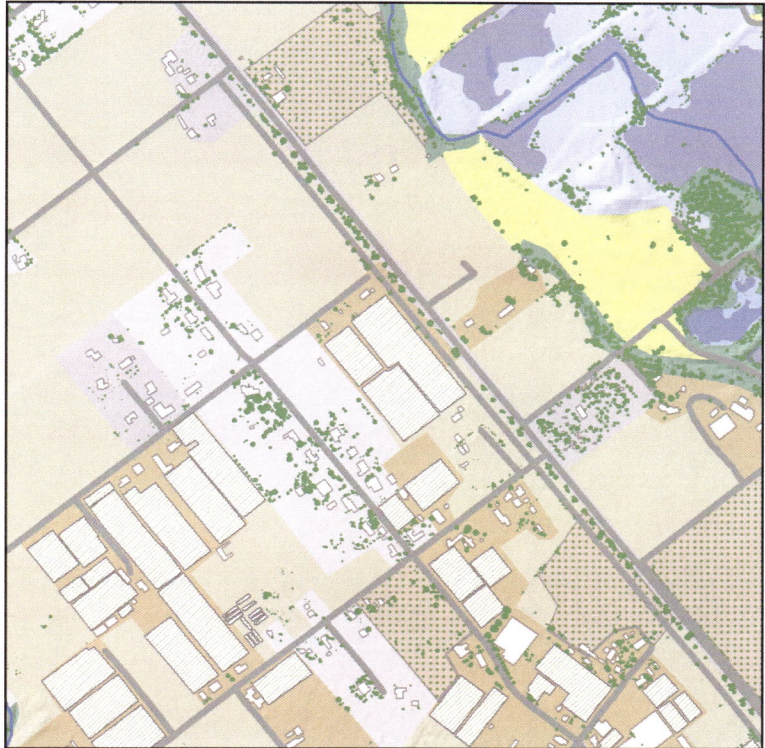

2070

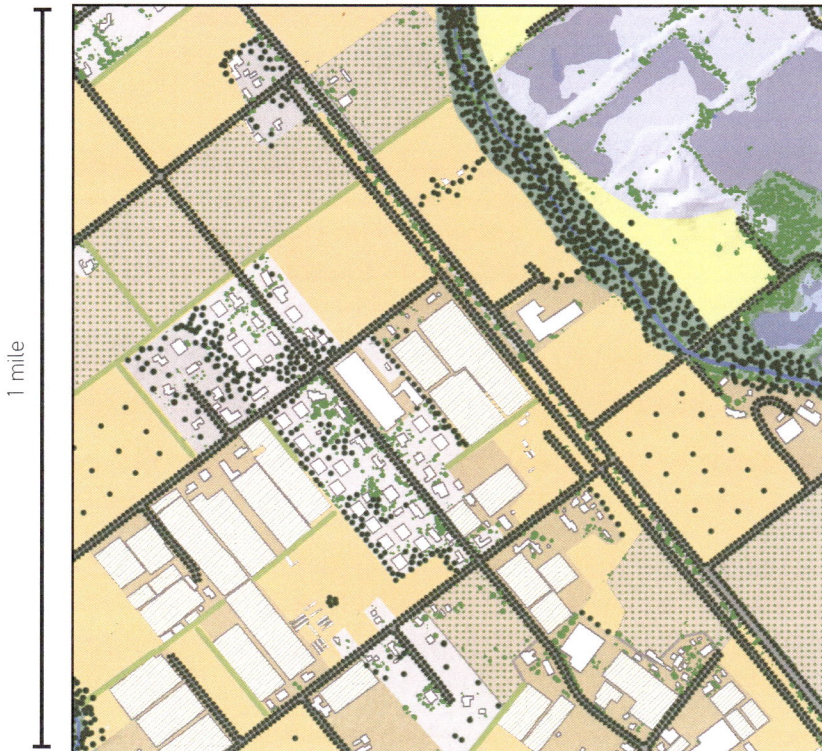

Legend

- Row crop, hay, and pasture
- Agriculture developed
- Climate-smart agriculture
- Orchard
- Oak woodland
- Grassland
- Freshwater marsh
- Riparian woodland
- Water
- Hedgerow
- Stream
- ● Existing tree
- ● New tree
- Building
- Greenhouse
- Rural residential
- Road

2070

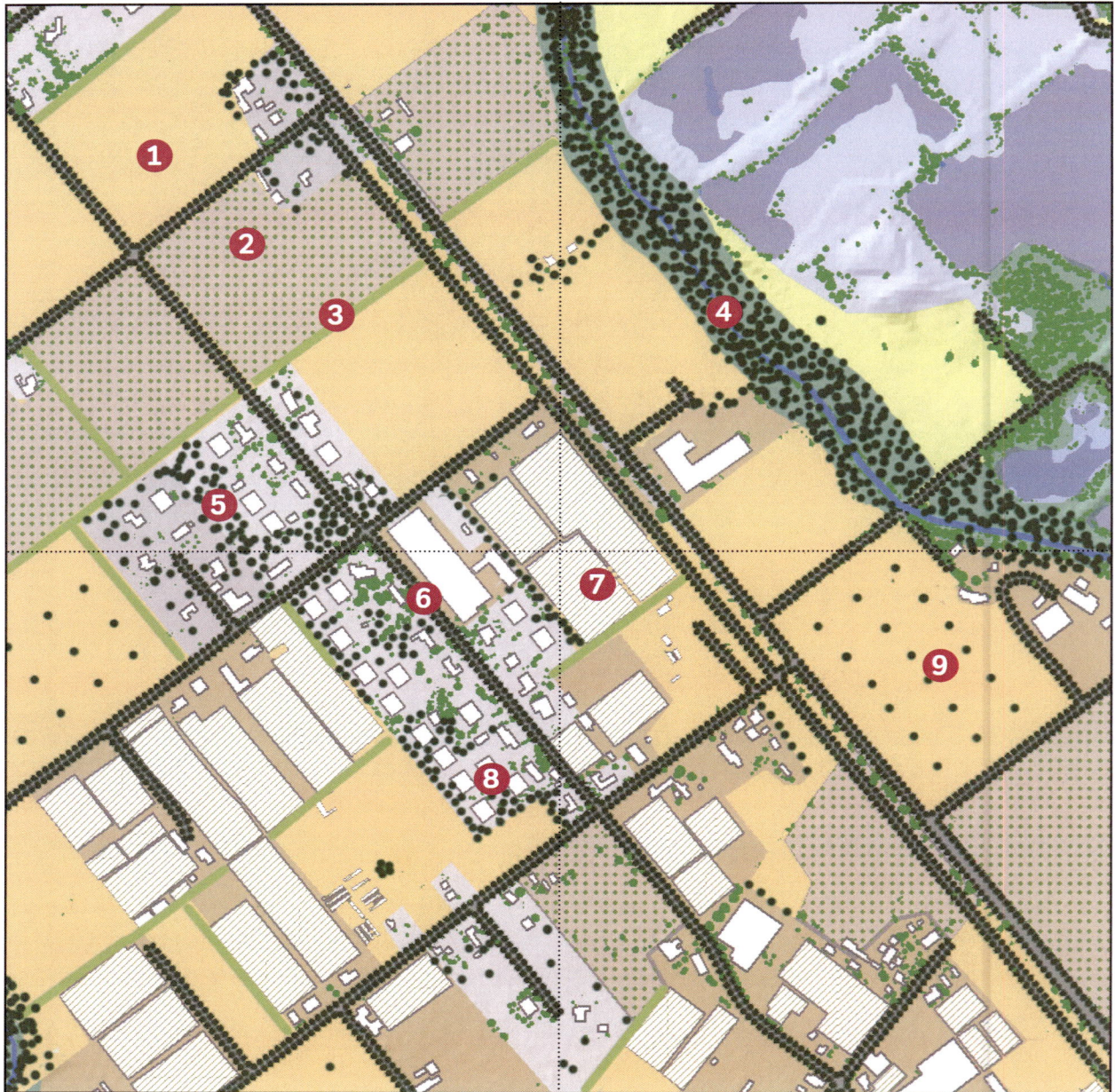

0.5 miles

Legend

	Row crop, hay, and pasture		Grassland	— Stream		Rural residential
	Agriculture developed		Freshwater marsh	● Existing tree	—	Road
	Climate-smart agriculture		Riparian woodland	● New tree		
	Orchard		Water	Building		
	Oak woodland		Hedgerow	Greenhouse		

Benefit quantification

- Climate smart agricultural practices, such as compost application, mulching, and hedgerow planting improve carbon sequestration, drought-tolerance, pest management, and more.
- Integrating hedgerows and trees in fields, planting riparian buffers, and installing wildlife crossings facilitate wildife movement across the valley floor, connecting populations.
- Preserving local food production reduces carbon costs from food distribution and preserves an agrarian culture.

Landcover Change. Riparian restoration leads to a net loss of 4 acres of agricultural land. However, it increases carbon sequestration and storage, improves water quality, and reduces flooding. Planting hedgerows and street trees and transitioning from row crops to orchards also bolster ecological connectivity.

Landcover Type	Existing	Change	Total	Unit
Orchard	39	37	76	ac
Row Crop, Hay, Pasture	203	-41	162	ac
Agriculture Developed	114	0	114	ac
Riparian Woodland	22	28	50	ac
Hedgerows	0	9,607	9,607	ft
Total Trees	735	1,271	2,006	

Approximate carbon sequestration and greenhouse gas emission reductions and payments associated with selected conservation practices. Climate-smart practices in agricultural land result in 1,205 tons of carbon sequestered annually or the equivalent of retiring 261 passenger vehicles. Altogether **landowners could receive up to $2.3 million in payments a year** from the California Department of Food and Agriculture's Healthy Soils Program for adopting these conservation practices for 5 years.

Conservation Practice	Area	Total CO_2 Equivalent (tons)	Estimated Payment
Compost Application	242 Acres	1,090	$290,400
Mulching	242 Acres	60	$1,806,390
Hedgerow Planting	9,607 Feet	18	$103,947
Riparian Buffer	21 Acres	42	$90,547
Total		1,205	$2,291,284

Legend (continued from previous page)

1. Compost, mulching, reduce till and other climate-smart practices
2. Cover crops and transition to other crops
3. Hedgerows and other pollinator corridors
4. Restoration of historic creek alignment and riparian woodland
5. Re-oaking in residential areas and along streets
6. Reduced speed limits to lower wildlife mortality rates
7. Preservation of profitable farming models
8. New farmworker housing in existing rural residential parcel
9. Re-oaking interspersed in farmland

Planting 1,271 additional trees increases carbon storage from 10,627 to 13,306 tons.
Trees can be added in restored riparian areas, along streets, in residential yards and parks, and in agricultural fields. Estimated value based on soil and above-ground carbon storage using NRCS/SSURGO, CREEC, and iTree Canopy. For more information on methodology refer to Appendix 1.

Cultivated land - carbon storage (tons)

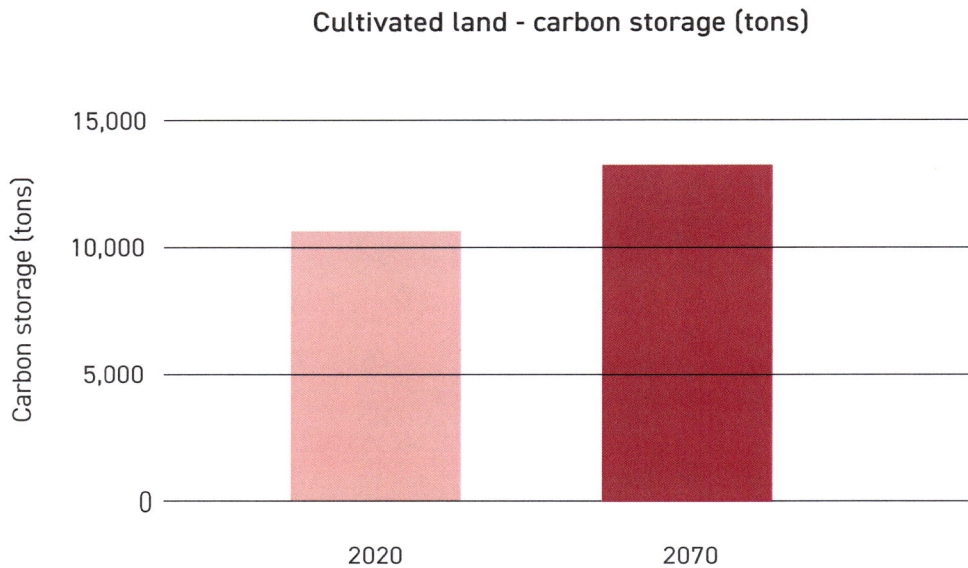

Planning and policy approaches for implementation

Zoning: Consider downzoning rural residential and office park areas that have not yet been developed. Create general plan overlays for climate resiliency, agricultural land preservation, and/or habitat connectivity based on ecosystem service valuation. Consider consolidating rural residential, agricultural facilities, and greenhouses to increase landscape connectivity for wildlife.

Buyout programs: Develop programs to buy back private land at market rates. Strategic buyouts of parcels located in existing and projected floodplains can help prevent damages and restore ecosystem function. Federal programs (e.g., Flood Mitigation Assistance Program and Hazard Mitigation Assistance grants), and state and local bonds, can help fund the strategic acquisition of land to meet resilience and habitat goals.

Exchange markets: Create transferable development rights programs or mitigation acknowledgement programs to discourage intentional fallowing of agricultural land. Develop an agriculture enterprise program that is based on low-wildlife-conflict crops to reduce the need for biocontrol. Pay farmers for adopting climate-smart management practices that store and sequester carbon. Reduce barriers and streamline application process for payment for ecosystem service programs.

Transportation: Reduce speed limits to reduce and slow traffic along roads to allow for movement of slower-moving agricultural equipment and reduce road mortality for wildlife. Build wildlife crossing infrastructure outside of roadways to create wildlife corridors that link populations in the Diablo Range and Santa Cruz Mountains.

Innovation: Fund adaptation studies to help landowners understand best management practices. Build an agricultural incubator to test innovative land management practices and quantify benefits.

Incentives: Develop agricultural conservation easements and other voluntary financial incentives for landowners, like the County's Agricultural Resilience Incentives grant program, to protect open space and enhance ecological connectivity. Offer funding to help farmers transition to more climate-appropriate crops.

Development: Institute a cap on development to limit the expansion or creation of facilities, contractor yards, and nurseries in prime farmland. Preserve, and when possible build more, farmworker housing in existing developed areas. Ensure that there is a diversity of housing affordability available but discourage new development in prime agricultural land. Build local storage, processing, and distribution facilities in existing developed areas to promote regional food resilience and reduce the carbon footprint of food production and distribution.

Rural & Open Space Scenario

A patchwork of land uses characterizes the rural & open space place-type. Open space here is not permanently protected and interfaces with agricultural, residential, and commercial land uses. Therefore, a variety of nature-based solutions are appropriate in this place-type to target the different opportunities and barriers of each land use. Like cultivated lands, these areas also face significant development pressure from the City of San José to the north and Morgan Hill and Coyote to the south. Policies such as conservation easements and zoning overlays are necessary to ensure adequate protection and management of open space in peri-urban areas.

Rural & open space is the most common place-type in Coyote Valley, totaling 120 cells (19,200 acres). 38% of these cells include a creek or tributary. Overall, land cover in these cells is predominantly composed of annual grasslands (25%), serpentine grasslands (19%) and grain, row crop, hay and pasture (11%). Tree canopy cover varies considerably (0-35%), due to the diversity of habitat types and soil conditions. Cells in the foothills of the Santa Cruz Mountains are characterized by mixed oak woodlands and in the Diablo Range by serpentine grasslands.

Key strategies for integrating nature-based solutions

Riparian corridors and buffers: Expand riparian buffers and floodplains to make room for natural processes (especially beneficial flooding), increase carbon sequestration, and capture pollutants.

Open space preservation: Allow for and increase groundwater recharge through open space preservation, floodplain restoration, and hillside capture. Protect existing habitat (open space with natural vegetation communities).

Prescribed grazing: Manage the intensity, frequency, timing, duration, and distribution of grazing to reduce erosion, recharge groundwater, control invasive plant species, and manage fire fuel loads.

Compost application: Apply compost on grasslands and farmlands to increase carbon sequestration, water infiltration potential, and plant and soil health, and to reduce erosion. Compost should only be applied in areas that will not impact native species.

Bioretention systems: Construct a treatment wetland at the confluence of two creeks to slow down and treat flows.

Wildlife corridors: Build overpass/underpass crossings along high use connections to reduce road mortality of wildlife.

Hedgerows: Establish rows of shrubs, grasses, or other perennial vegetation between fields and greenhouses to provide nesting, forage, and shelter for pollinators, mammals, birds, reptiles, and amphibians. Hedgerows can also prevent wind-driven erosion, capture air pollutants, and sequester and store carbon. Coordinate hedgerow locations between different parcels to create continuous wildlife corridors.

Management: Adopt wildlife-friendly farming practices (e.g., adding raptor boxes, reducing use of pesticides and fertilizers) that also improve water quality and capture particulate matter.

2020

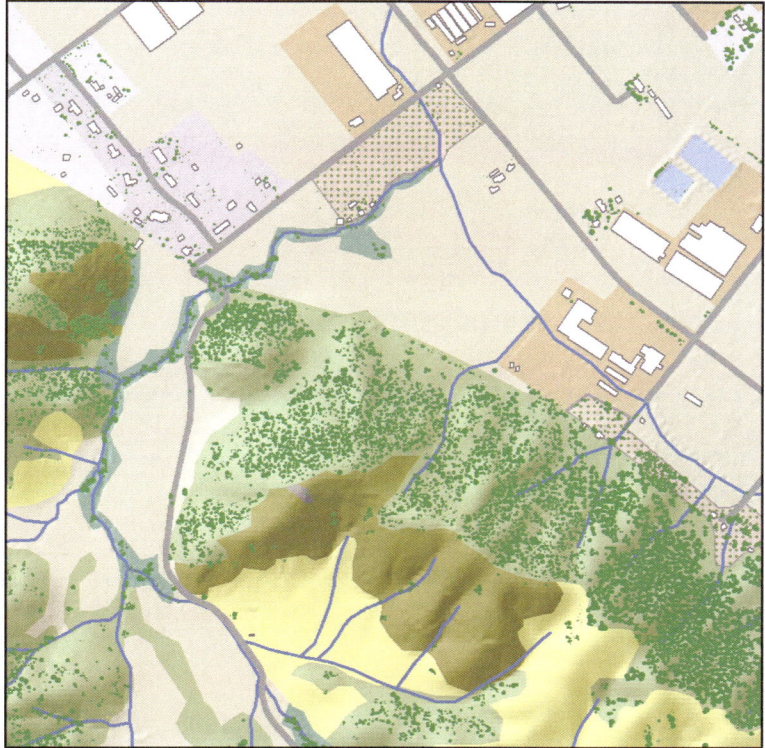

1 mile

2070

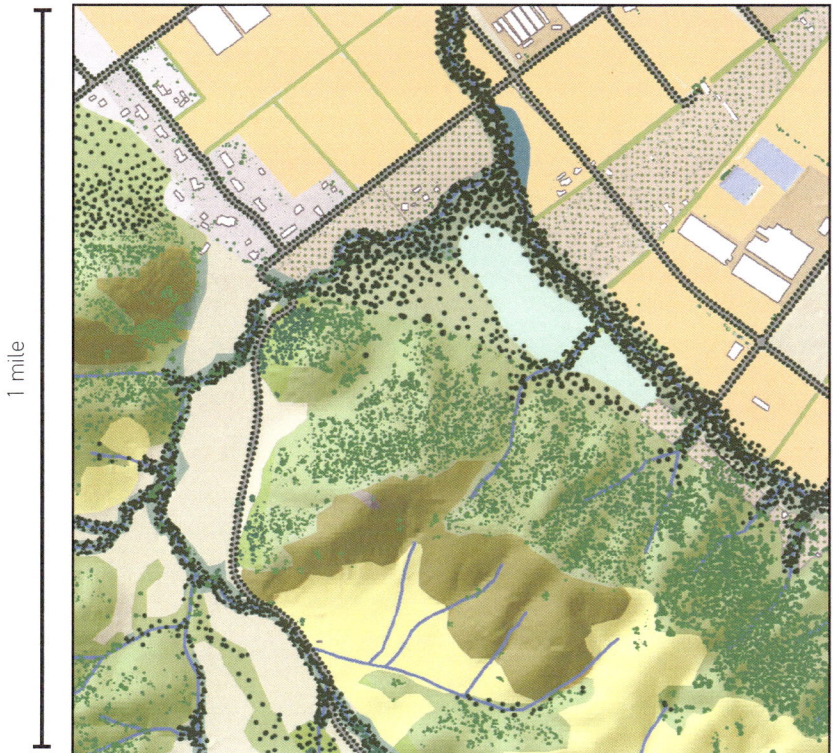

1 mile

Legend

- Row crop, hay, and pasture
- Agriculture developed
- Climate-smart agriculture
- Orchard
- Rural residential
- Wet meadow
- Oak woodland
- Chaparral
- Grassland
- Riparian woodland
- Hedgerow
- Water
- Stream
- ● Existing tree
- ● New tree
- Building
- Road

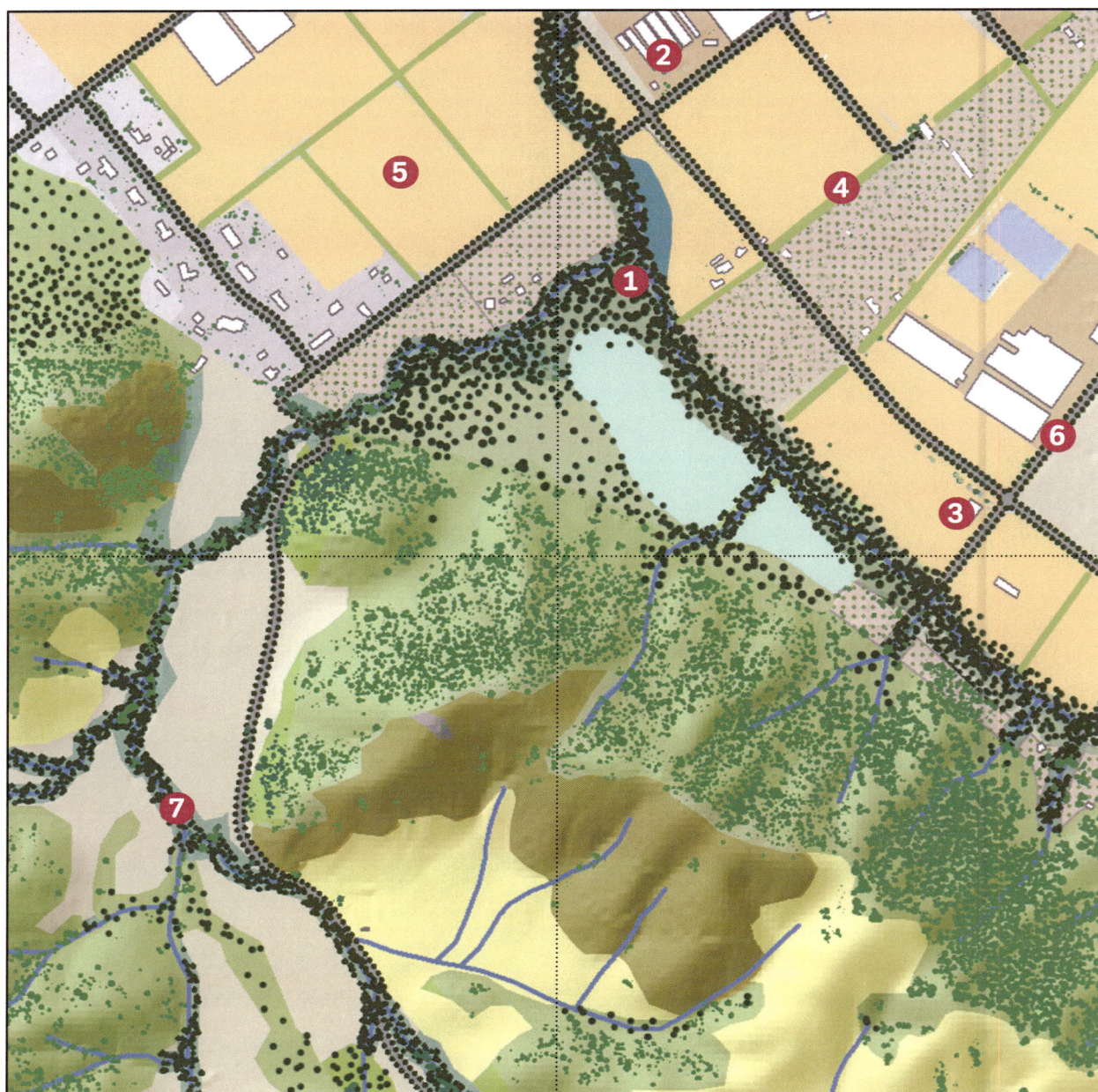

0.5 miles

Legend

	Row crop, hay, and pasture		Riparian woodland
	Agriculture developed		Hedgerow
	Climate-smart agriculture		Water
	Orchard	—	Stream
	Rural residential	●	Existing tree
	Wet meadow	●	New tree
	Oak woodland		Building
	Chaparral		Road
	Grassland/oak savanna		

1 Habitat and floodplain restoration at confluence on former agricultural land

2 Relocated mushroom facility away from creek

3 Ecology education center

4 Hedgerows and pollinator corridors

5 Agricultural climate-smart practices

6 Street tree canopy corridors

7 Riparian corridor restoration

Benefit quantification

- Rural & Open Spaces have the greatest potential for carbon storage out of the six place-types studied in this report.
- Oak and riparian woodland restoration improve habitat quality and greatly increase carbon sequestration and storage.
- Creek restoration is vital to landscape-scale flood control strategies that benefit urban areas downstream.

Landcover Change. The Rural & Open Space place-type presents an opportunity for large-scale habitat restoration and enhancement, as well as reforestation.

Landcover & Features		Existing	Change	Total	Unit
Agriculture	Orchard	21	26	47	ac
	Row Crop, Hay, Pasture	193	-70	123	ac
	Agriculture Developed	44	-26	18	ac
Habitat type	Oak Woodland & Mixed Forest	290	29	319	ac
	Riparian	15	27	42	ac
	Wet Meadow		11	11	ac
Features	Hedgerows		9,492	9,492	ft
	Street Trees		225	225	
	Total Trees	2,178	1,145	3,323	

Climate-smart practices in agricultural land result in 871 tons of carbon sequestered annually, or the equivalent of retiring 189 passenger vehicles. Altogether, **landowners could receive up to $1.7 million in payments per year** for adopting these conservation practices for 5 years.

Conservation Practice	Area	Total CO_2-Equivalent (tons)	Estimated Payment
Compost Application	170 Acres	760	$204,000
Mulching	170 Acres	41	$1,268,951
Hedgerow Planting	9492 Linear Feet	18	$102,703
Riparian Buffer	27 Acres	53	$116,418
Total		**871**	**$1,692,073**

Planting 1,145 additional trees in restored riparian areas, along streets, in residential areas, and in agricultural fields significantly increases the study cells' carbon storage from 14,383 to 16,373 tons. Estimated value based on soil and above-ground carbon storage using NRCS/SSURGO, CREEC, and iTree Canopy. For more information on methodology refer to Appendix 1.

Rural & Open Space- carbon storage (tons)

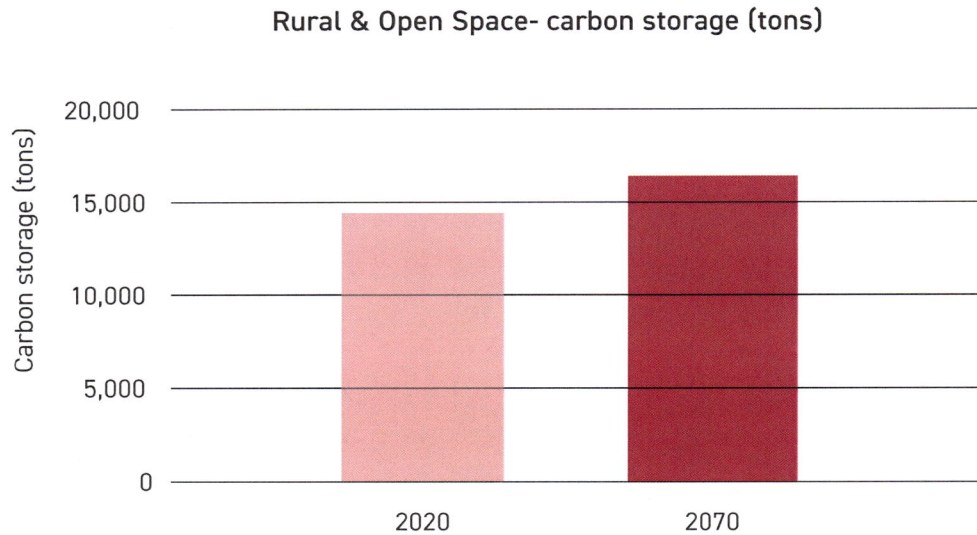

Planning and policy approaches for implementation

Zoning: Consider downzoning rural residential and office park areas that have not yet been developed. Create general plan overlays for climate resiliency, agricultural land preservation, and/or habitat connectivity based on ecosystem service valuation. Consider consolidating rural residential, agricultural facilities, and greenhouses in one zone to increase landscape connectivity for wildlife.

Incentives: Develop agricultural conservation easements and other voluntary financial incentives for landowners to protect open space and enhance ecological connectivity. Offer funding to help farmers transition to more climate-appropriate crops.

Development: Institute a cap on development to limit the expansion or creation of facilities, including mushroom growing facilities, contractor yards, and nurseries, in prime farmland. Retire or relocate facilities along creeks.

Exchange markets: Create transferable development rights programs or mitigation acknowledgement programs to discourage intentional fallowing of agricultural land. Develop an agriculture enterprise program that is based on low-wildlife-conflict crops to reduce the need for biocontrol. Pay farmers for adopting climate-smart management practices that store and sequester carbon. Reduce barriers and streamline application process for payment for ecosystem service programs.

Transportation: Reduce speed limits to slow traffic along roads and reduce road mortality for wildlife. Build wildlife crossing infrastructure outside of roadways to create wildlife corridors that link populations in the Diablo Range and Santa Cruz Mountains.

Innovation: Fund adaptation studies to help landowners understand best management practices. Build an agricultural incubator to test innovative land management practices and quantify benefits. Create a research program to pilot carbon farming and holistic grazing practices to address impacts of focused grazing.

Outreach and education: Add an education center to provide opportunities for people to learn first-hand the value of natural infrastructure and to promote environmental stewardship. Launch extensive education and outreach programs in preparation for ballot measures to gain support from landowners. Build a trail network to increase public access to outdoor recreation opportunities and promote land stewardship.

Parks & Protected Areas Scenario

Open space preservation increases the potential to adopt multi-benefit climate-smart land management practices, restore and protect valuable habitat types at a large scale, and expand outdoor public recreation. Strategic land acquisitions can help protect critical wildlife linkages, support regional populations sensitive to urban environments, and provide dedicated areas for species to migrate in response to a changing climate. This place-type is typically expected to be managed in a natural condition in perpetuity. Parks & protected areas are well-suited to pilot and monitor nature-based solutions that can be expanded and scaled into surrounding areas and other place-types.

There are 37 parks & protected areas cells (5,920 acres) in Coyote Valley, 51% of these are along a creek or tributary. Cells are mostly located on the eastern side of the valley in the foothills of the Diablo Range, which tends to be drier. The three most common land cover types are serpentine grasslands (38%), annual grasslands (16%), and urban/industrial development (8%). Tree canopy cover varies considerably (0-23%) due to the diversity of habitat types and soil conditions found in parks & protected areas, some of which do not support tree species.

Key strategies for integrating nature-based solutions

Wetland restoration: Restore seasonal and perennial wetlands that can improve water quality, buffer flooding downstream, replenish groundwater, sequester carbon, fuel the aquatic food web, and provide regionally rare habitat for resident and migratory wildlife. Remove earthen dams and fill in engineered drainage channels to recreate more natural hydrology. Reconnect creeks to their natural floodplains.

Riparian corridors and buffers: Expand riparian buffers and floodplains to make room for natural processes (especially beneficial flooding), increase carbon sequestration, and capture pollutants.

Habitat restoration: Restore willow grove and oak savanna habitat in current dry-land agricultural areas. Create more transitional habitat zones between serpentine grassland and seasonal wetland.

Prescribed grazing: Manage the intensity, frequency, timing, duration, and distribution of grazing to reduce erosion, recharge groundwater, control invasive plant species, and manage fire fuel loads.

2020

2070

Legend

- Row crop, hay, and pasture
- Seasonal wetland
- Perennial wetland
- Riparian
- Willow grove
- Oak mix
- Wet meadow
- Grassland
- Rural developed
- Water
- Stream
- Existing tree
- New tree
- Building
- Road

2070

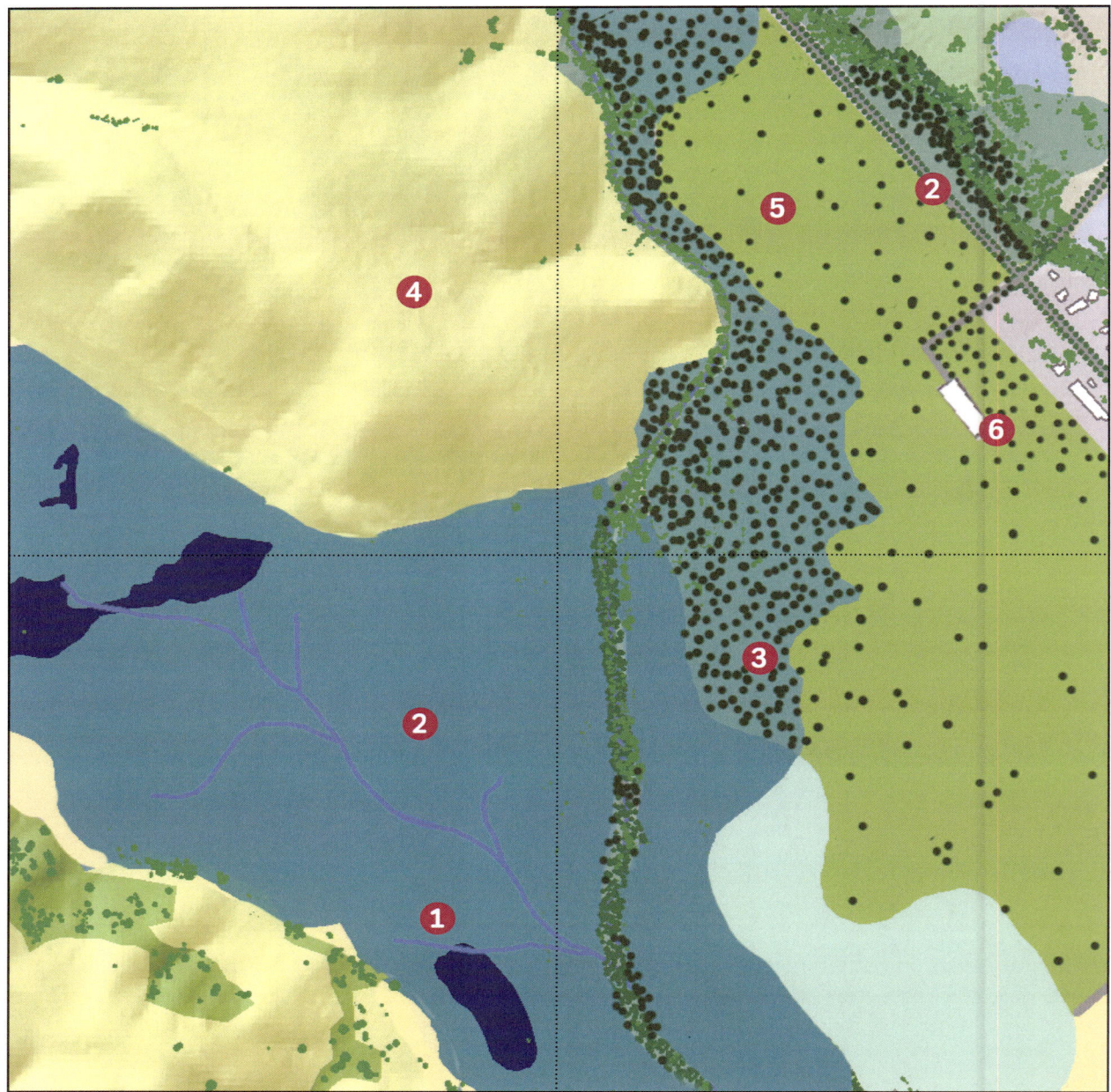

0.5 mile

Legend

▨ Seasonal wetland		▨ Rural residential		
▨ Perennial wetland		▨ Water		
▨ Riparian		— Stream		
▨ Willow grove		● Existing tree		
▨ Oak woodland		● New tree		
▨ Chaparral		☐ Building		
▨ Grassland		— Road		

1 Regionally-rare seasonal wetland restored and creeks reconnected to natural floodplains in current dry-land agriculture

2 Roads retired or elevated to improve ecological and hydrological connectivity

3 Restored willow grove

4 Prescribed grazing in grasslancs

5 Oak savanna in former industrial land

6 Ecology field station and educational center

Benefit quantification

- Restoring wetlands in the upper watershed has the greatest potential for reducing flooding downstream in urban areas.
- Unique habitat types, such as seasonal wetlands and serpentine grasslands support locally endemic species that are rare or threatened.
- Protecting and restoring land in the valley floor facilitates wildlife movement between the Santa Cruz Mountains and Diablo Range. Connectivity benefits species with larger home ranges, enables migration in response to changes in climatic condition, and provides an escape from wildfire events in the mountain ranges.
- Conservation and restoration of habitats in protected areas can provide the greatest support to regional biodiversity.

Landcover Change. Grasslands, oak savannas, wet meadows, seasonal and perennial wetlands, riparian woodlands, and forest groves are restored in this scenario, totaling **390 acres of habitat restoration.**

Landcover & Trees		Existing	Change	Total	Unit
Habitat type	Grassland	183	3	186	ac
	Oak Savanna	14	119	133	ac
	Wet Meadow		32	32	ac
	Seasonal Wetland	24	174	198	ac
	Perennial Wetland		10	10	ac
	Riparian	23	17	40	ac
	Willow Grove		35	35	ac
Trees	Street Trees		57	57	
	Total Trees	509	977	1,486	

Water Storage. Assuming an average depth of 18", the restored seasonal and perennial wetlands in this study cell could hold 13,590,000 cubic feet of water or the equivalent of 154 Olympic-size swimming pools.

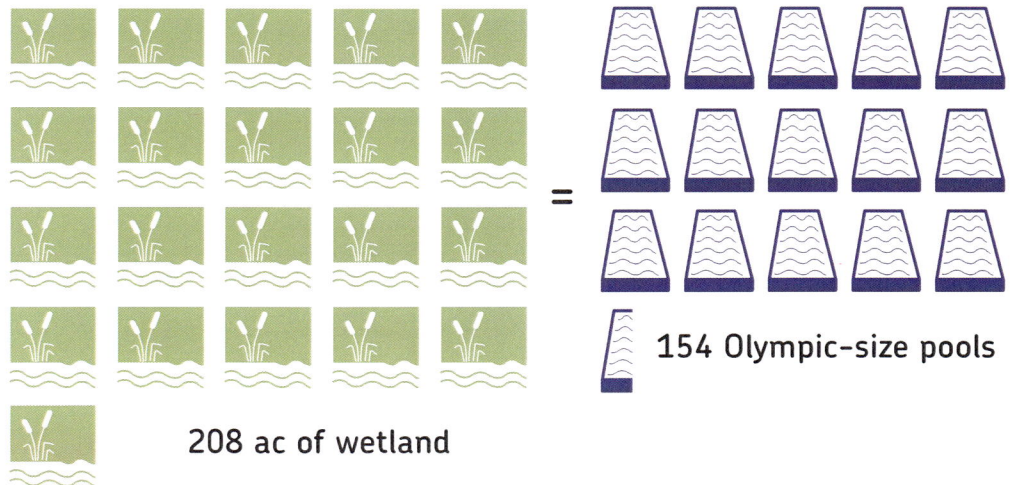

= 154 Olympic-size pools

208 ac of wetland

The restoration of oak savanna, willow grove, and riparian woodland ecosystems increases the study cells' soils and above-ground carbon storage from 10,302 to 12,812 tons. Estimated value based on soil and above-ground carbon storage using NRCS/SSURGO, CREEC, and iTree Canopy. For more information on methodology refer to Appendix 1.

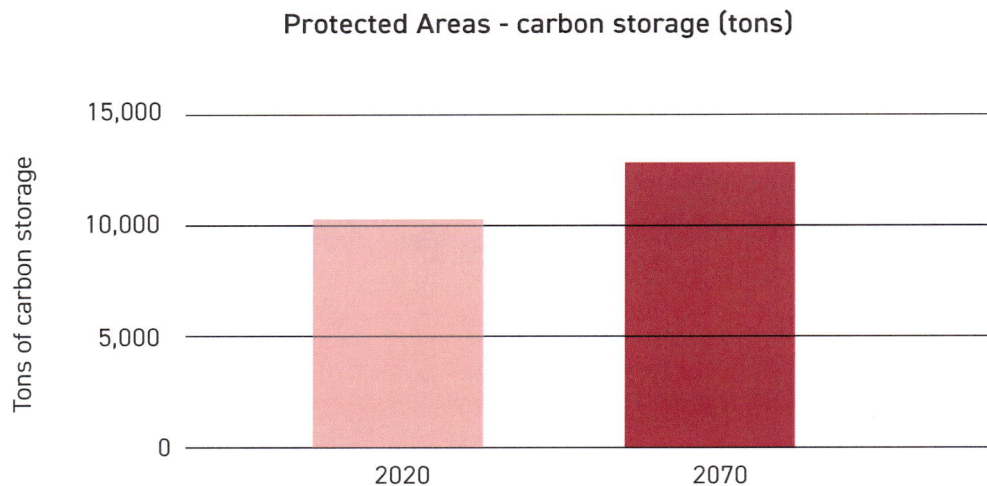

Protected Areas - carbon storage (tons)

Laguna Seca. Photograph by J. Plotsky, courtesy of OSA.

Planning and policy approaches for implementation

Transportation: Elevate or retire roads at critical wildlife links to increase ecological and hydrological connectivity. Build wildlife crossing infrastructure outside of roadways to create wildlife corridors that link populations in the Diablo Range and Santa Cruz Mountains. Improve public transportation to increase car-free access and promote park visitation (e.g., use rail line as an access point or set up a shuttle system).

Master plan: Develop a master plan vision for the phased restoration of newly protected land with the help of an interdisciplinary group of experts.

Outreach and education: Add an education center to provide opportunities for people to learn first-hand the value of natural infrastructure and to promote environmental stewardship. Set up a robust community outreach strategy to incorporate stakeholder feedback early in the process. Build a trail network to increase public access to outdoor recreation opportunities and promote land stewardship.

Development: Remove industrial land uses and increase building setbacks along creek to make room for restoration. Retire natural gas plant to reduce fossil-fuel derived electricity.

Exchange markets: Create transferable development rights programs or mitigation acknowledgement programs to compensate, if possible, loss of agriculture in this cell with permitted or assisted agricultural intensification elsewhere. Use carbon and mitigation credit markets to finance improvements.

Zoning: Create habitat and ecosystem-service zoning overlays to achieve creek setbacks and habitat restoration on private land.

Incentives: Create an incentive-based program for landowners and leaseholders to manage grazing in grasslands.

Infrastructure: Bury power lines to reduce fire hazards.

▲ Laguna Seca. Photograph by J. Plotsky, courtesy of OSA.

Across the Urban-to-Rural Gradient

Climate change plans will be most effective when thinking across scales and from the urban-to-rural gradient. Political jurisdictions usually do not follow the boundaries of watersheds and other natural systems. Adaptation plans will be most effective if they respond to the heterogeneity of place-types that make up our cities and counties. The land use mix, surface permeability, housing density, and job density characteristics of each place-type influence which nature-based solutions would be best suited and most impactful.

Urban and rural areas will need to be retrofitted to better accommodate nature and take advantage of ecosystem services. The place-type framework identifies landscape typologies that can help craft general strategies and policies for climate adaptation that are sensitive to the diversity and complexity of urban and rural conditions. However, each cell is unique and will require site and neighborhood-scale considerations. Solutions will vary depending on land use, built form, lot size, functional needs, infrastructure age, equity, climate risks, and a variety of site-specific conditions. For example, the construction of tall buildings in cities presents an opportunity to integrate vertical vegetated elements such as green roofs, terraces, and walls that aren't as necessary, or appropriate, in low-rise neighborhoods with more available open space. Some of these measures might be more likely deployed in certain place-types but should not be excluded from others. Mulching, composting, and prescribed grazing are part of a suite of climate-smart agricultural practices that can be extended to urban farms and large parks. The

general suitability of different measures is summarized in Figure 1 by place-type based on the previous scenarios.

Restoring natural elements and processes in one place-type also has the potential of benefiting others. Building parks, greenways, and a robust urban forest can increase biodiversity support and the ecological permeability of urban areas, enhancing the value of habitat protection and restoration in adjacent rural areas. Similarly, flood control and stormwater management is most effective when planned at the watershed-scale. Floodplain preservation and groundwater recharge upstream can be complemented downstream by stormwater capture and green infrastructure projects. Creeks and rivers connect the Baylands to hillside wildlife populations. Doing restoration along the length of Coyote Creek, from urban neighborhoods to parks & protected areas, reinforces the ecological connectivity between place-types and historically disconnected aquatic and terrestrial habitat types.

Figure 1: Suitability of nature-based solution by place-type.

	Urban Neighbor-hoods	Office Parks	Cul-de-Sac Suburb	Cultivated Land	Rural & Open Space	Parks & Protected Areas
Parks	H	H	H		H	H
Riparian Corridor Restoration & Buffers	H	H	H	H	H	H
Greenways	H	H	H	S		
Bioretention areas	H	H	H	H		
Yard improvements	H	H	H	H	S	
Street trees	H	H	H	H	H	
Green roofs	H	H	S	S	S	
Green terraces	H	H				
Green walls	H	H				
Hedgerows				H	H	
Re-Oaking	H	H	H	H	H	H
Wetland restoration		H		H	H	H
Cover cropping				H	H	S
Compost application				H	H	H
Reduce tilling				H	H	S
Mulching				H	H	S
Prescribed grazing				H	H	H

H = High suitability
S = Some suitability

Biker on Guadalupe Creek trail. Photograph by Sergio Ruiz, courtesy of SPUR

Across the Urban-to-Rural Gradient *(continued)*

Nature in cities directly benefits the greatest number of people and makes density more livable. However, planners and architects must incorporate ecological planning early to prevent development and transportation from taking up strategic ground-level space that can be dedicated to nature-based solutions. Reducing building footprints and transitioning away from car-oriented mobility is key in making room for nature. The future place-type scenarios shown in this chapter identified widening riparian corridors, creating publicly accessible greenspaces, expanding the urban forest, and integrating bioretention features as the highest priority interventions.

Urban ⬅ ·······································

Urban Neighborhoods | Office Parks | Cul-de-Sac Suburbs

This place-type **outperformed Office Parks and Cul-de-Sac Suburbs** in carbon sequestration, carbon storage, avoided runoff, and air pollution removal, despite having the highest building area (around 40%) thanks to its **existing mature canopy and an ambitious tree planting strategy**. **Bioretention features along streets and in parks** were identified as high priority nature-based solutions to regulate microclimates and reduce runoff in this place-type given that this cell has the highest impervious cover (about 71%).

Transitioning from surface parking lots and single-story buildings to district parking structures and mid to high-rise buildings freed up **37 acres of land** (or 28% of the cell) **for greenspace** creation. Structures that were located within the existing floodplain were relocated closer to the transit stop allowing for the restoration of the riparian corridor and a big reduction in flood risk. These large greenspaces can act as **ecosystem service hubs within urban areas** and serve a wide variety of uses, including community gardens, detention basins, wildlife habitat, and immersive natural areas.

The future Cul-de-Sac Suburb scenario provided the **least ecosystem services out of the six place-type cells** and absorbed the lowest number of housing units and jobs out of the urban place-types. However, Cul-de-Sac Suburbs **are the most common**, totaling 252 cells in San José. Investments in riparian restoration, green infrastructure, backyard improvements, and other nature-based solutions, **when aggregated, have the greatest potential to provide the greatest benefits.**

Rural areas have more open space and therefore the greatest opportunity for large-scale interventions. Habitat restoration and climate-smart conservation practices can enhance ecosystem service provision outside of cities, where it is already relatively higher. Creating and protecting habitat patches and wildlife corridors in rural areas is essential for supporting species that are sensitive to urban conditions and species with larger home ranges. These measures benefit humans as well. Nature-based solutions can can simultaneously support biodiversity and serve as regional stormwater and decarbonization infrastructure.

Rural

Cultivated Land

Farmers can receive **money from ecosystem-service payment programs for adopting climate-smart practices**. These measures not only improve the ecological value of their fields and orchards but also contribute to **lowering water demand, increasing carbon sequestration, reducing air pollution, and buffering flooding**. Protecting local agriculture can help reduce the carbon footprint of our food and can be achieved while restoring ecosystem functions.

Rural & Open Space

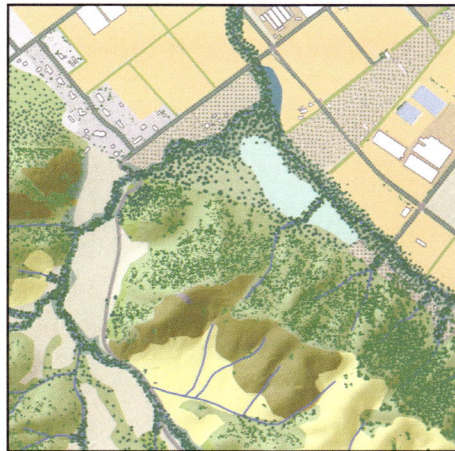

Many types of nature-based solutions are suited to this place-type given the patchwork of land uses. This also means that Rural & Open Spaces confer many benefits, from the **most carbon storage and carbon sequestration** to improving agricultural productivity. **Restoring oak woodlands and riparian forests, and installing hedgerows** are some of the most impactful measures.

Parks & Protected Areas

The protected status of lands in this place-type lends itself to the **largest habitat restoration projects** and provides unique opportunities for recreation and outdoor education in natural ecosystems. Restoring seasonal and perennial wetlands in this cell was selected as the most important improvement by charrette participants. Not only would this rehabilitate a **regionally rare habitat type and critical wildlife link**, but it would also increase **groundwater recharge and water storage** that would benefit urban areas downstream.

Comparing Carbon Storage Potential

Rural areas store more carbon than their urban counterparts. This is consistent with findings presented in Beller et al. 2020, showing a 50% loss of tree carbon storage in Silicon Valley as a result of Euro-American landscape modification and urbanization. Carbon storage in the Rural & Open Space cell was the highest and increased from 3,596 to 4,093 tons. However, our scenario-based analysis also suggests that Urban Neighborhoods and Office Parks can more than quadruple above-ground carbon storage in the next fifty years, complementing the contribution of rural areas. Tree canopy cover is currently very low in urban areas and planting more trees can be one of the most effective ways to increase carbon storage and other ecosystem services. Despite integrating nature-based solutions, the Cul-de-Sac Suburbs cell had the lowest current and projected carbon storage values. This underscores the importance of protecting rural land and preventing further sprawl.

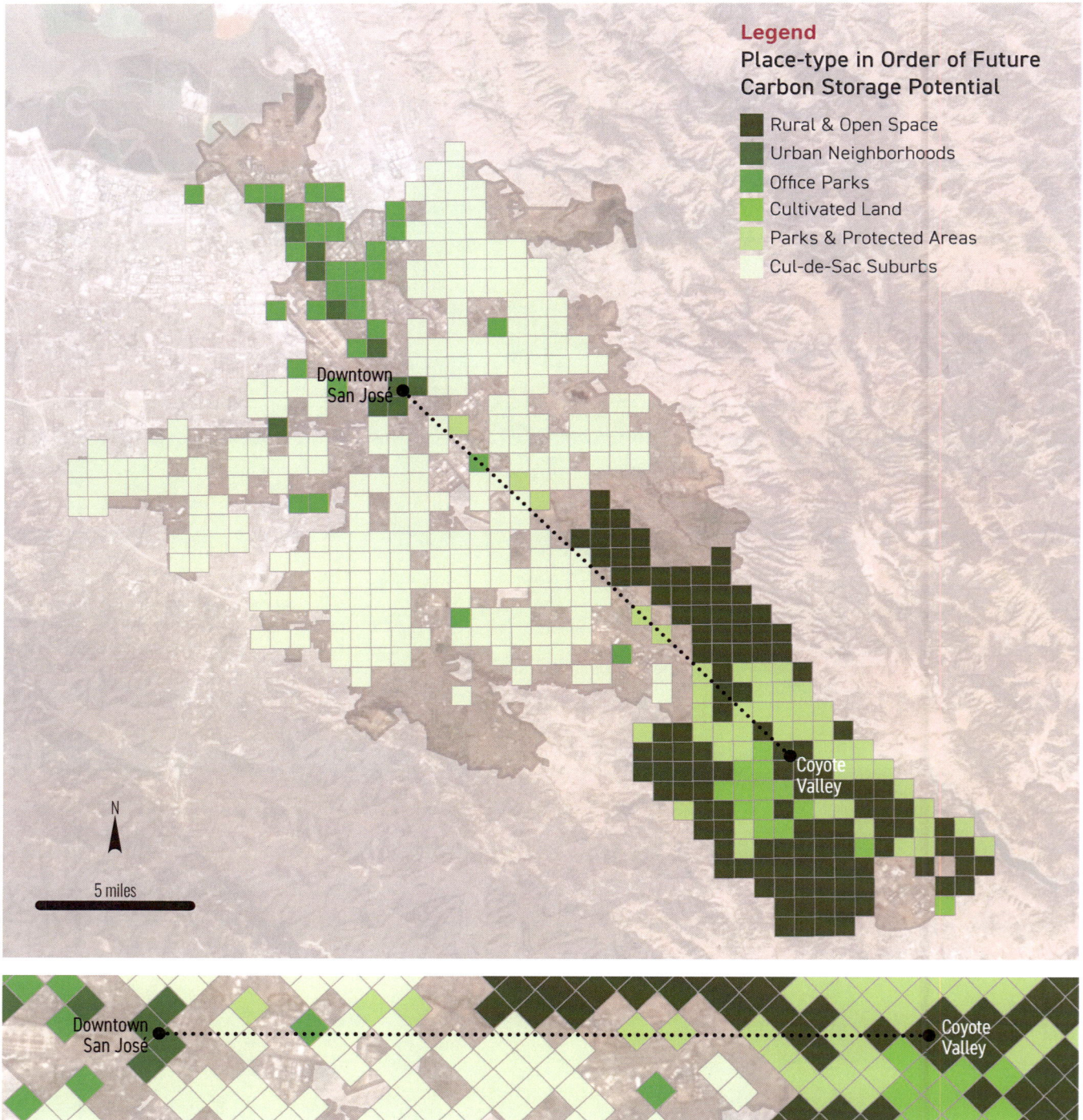

Legend

Place-type in Order of Future Carbon Storage Potential

- Rural & Open Space
- Urban Neighborhoods
- Office Parks
- Cultivated Land
- Parks & Protected Areas
- Cul-de-Sac Suburbs

We quantified values for a single grid cell of each place-type, but there is variation among cells of the same place-type. For example, the Parks & Protected Land cell in this study includes a regionally rare seasonal wetland, a unique condition that is not representative of opportunities in other cells with different habitat types, and most likely underestimates this place-type's potential for carbon storage. Aggregating benefits across San José and Coyote Valley would require further analysis on the variation of conditions and opportunities of each cell.

For the purposes of illustrating how all six place-types can be scaled-up to the city or county-scale, we multiplied our scenario-based values by the total number of cells of that place-type in our study area. We estimate that **deploying nature-based solutions in these place-types could increase the value of carbon storage by $117 million, totaling $262 million in the year 2070.** This conservative estimate demonstrates the immense benefits of restoring ecosystem function across the landscape and doesn't take into consideration the value of multiple co-benefits such as microclimate regulation, air pollution removal, water quality improvement, food production, flood control, and more.

Rural place-types currently store more carbon than urban place-types, but urban place-types have the potential to change that through aggressive reforestation plans. The estimated value of urban place-types shown below is based on above-ground carbon storage using iTree Eco. Estimated value of rural place-types is based on soil and above-ground carbon storage using NRCS/SSURGO, CREEC, and iTree Canopy. For more information on methodology refer to Appendix 1.

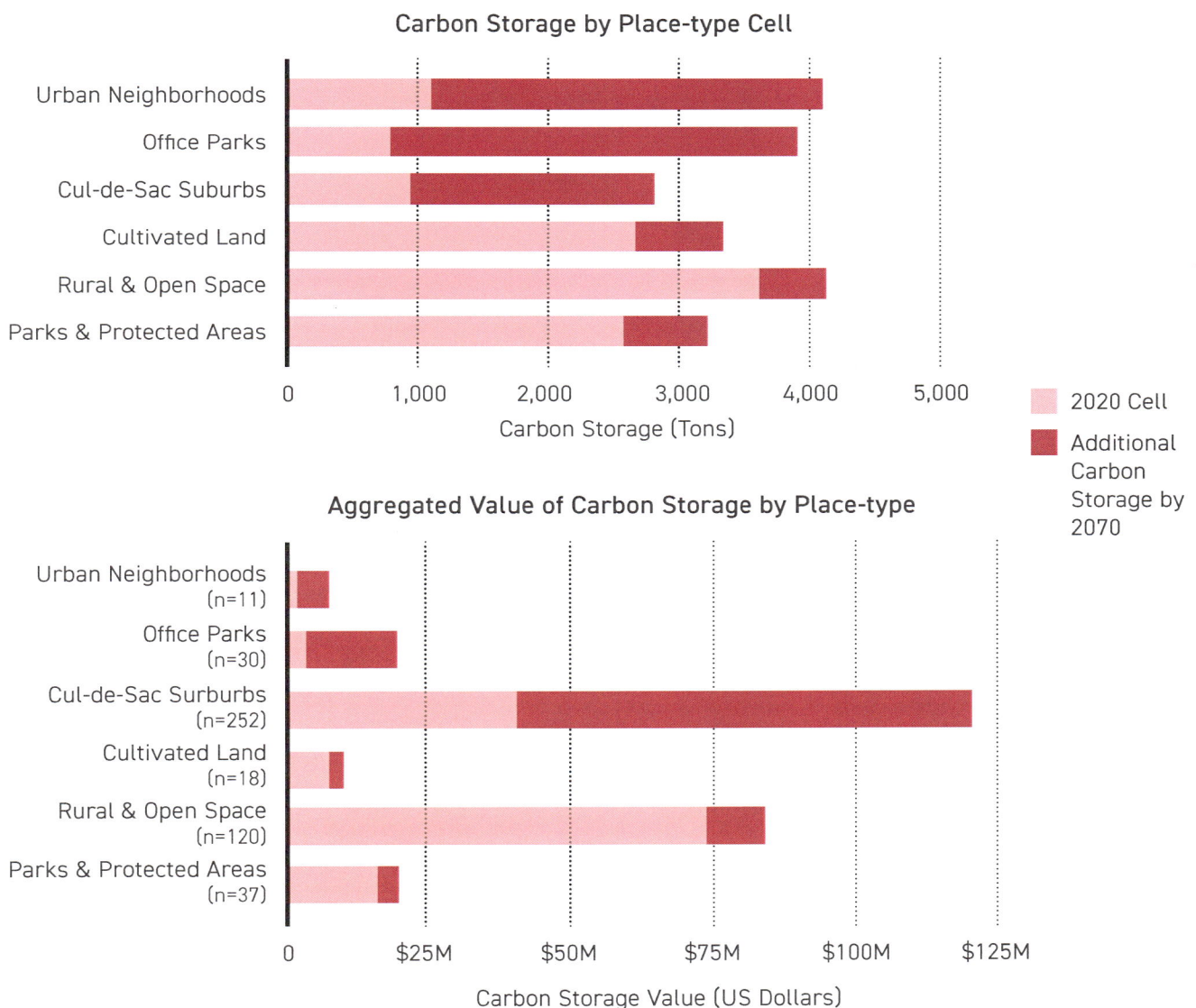

Carbon Storage by Place-type Cell

Legend:
- 2020 Cell
- Additional Carbon Storage by 2070

x-axis: Carbon Storage (Tons)

Aggregated Value of Carbon Storage by Place-type

x-axis: Carbon Storage Value (US Dollars)

4

TOWARD A MORE
RESILIENT
LANDSCAPE

Successfully deploying the climate adaptation strategies outlined in Chapter 3 across an urban-to-rural gradient will require extensive coordination among policymakers, landowners, and other stakeholder groups. The future scenarios for each place-type contain many ambitious nature-based interventions, yet various policy measures already exist to make them tenable. Interdisciplinary collaboration and landscape-scale planning are key to ensure benefits are optimized across sectors. Aligning policies across levels and agencies of government may increase the success of planning efforts to build community resilience with nature. The following section summarizes high-leverage themes from Chapter 3 and presents policies and strategies available for implementing them. The list of policies and programs below is not exhaustive, and determining the correct policy measures for a given locality will require stakeholder input.

Left top: Cyclist in San José. Photograph by Richard Masoner, courtesy of CC 2.0.
Left bottom: Green heron at Coyote Creek. Photograph by Alan Hack, courtesy of CC 2.0.

Waterways link urban and rural areas and present one of the greatest opportunities for habitat and ecosystem-service restoration across the urban-to-rural gradient.

Both in San José and Coyote Valley, buildings, infrastructure and agriculture have encroached on creeks and floodplains, making urban and rural areas more vulnerable to flood events and debilitating critical wildlife corridors. Preventing further encroachment into riparian areas and removing existing development near waterways can lower the risk of property losses during flood events and restore the ecological function of riparian areas. Creeks and rivers serve as vital ecological corridors, connecting the Baylands to urban nature and mountain ranges. Wider, forested riparian areas can also increase carbon sequestration, improve water quality, capture air pollutants, replenish aquifers, and offer recreational opportunities.

Conservation easements and strategic land acquisitions are tools for protecting and expanding riparian corridors and keeping development out of harm's way. Following repeated flood events, cities and counties can apply for the Federal Emergency Management Agency's Hazard Mitigation Grant Program to fund the strategic acquisition of properties in high risk areas. For example, in Texas, the Harris County Flood Control District developed a program to buy out homeowners in the ten-year floodplain to prevent future flood damages. However, innovations in flood insurance programs are necessary to relocate properties before flood events. Integrating **pre-approved and guaranteed buyouts** as an insurance benefit for homeowners in high risk zones could help reduce the social and economic impacts of flood events. Another incentive for riparian restoration and enhanced setbacks is the California Department of Food and Agriculture's Healthy Soils Program, a **payment for ecosystem services program** that facilitates the installation of riparian buffers in farmland that contribute to on-site and downstream resilience. Areas that are not directly impacted by flooding events can also help improve water quality and reduce peak flows by installing **bioretention systems and reducing impervious cover** to increase water infiltration.

Top: Coyote Creek at Tully Road, San José. Imagery courtesy of Google Earth.
Bottom: Riparian buffer, Courtesy of Bear Creek National Agroforestry Center.

Transportation improvements play a key role in building resilience at a landscape scale.

Upgrading transportation networks will be necessary in order to enable sustainable densification, reduce conflict with wildlife movement, improve hydrological connectivity, and build natural capital in the public right-of-way. Expanding public transportation networks to reach outdoor recreation areas, such as the newly acquired North Coyote Valley Conservation Area, can reduce traffic and carbon emissions while helping address existing inequities in park access.

Car-oriented planning has dedicated a significant amount of land to impervious road networks that exacerbate urban heat islands, worsen local air quality, raise flood risks, diminish urban walkability, and decrease the habitat area available for wildlife. Disincentivizing single-passenger vehicles, particularly fossil-fuel powered vehicles, will therefore play a crucial role in decarbonizing San José and Coyote Valley and mitigating climate risks. San José has the opportunity to free up valuable ground-level space for street trees, planting buffers, bioretention areas, spacious sidewalks, and bike lanes that improve the livability of urban areas by **eliminating parking construction minimums and reducing road lanes.** This approach has already been tested by cities, such as San Francisco, in an effort to promote walkability and reduce the cost of housing construction.

Top: First Street, San José. Photograph by Sergio Ruiz, courtesy of SPUR.
Bottom: Wayfinding signs, San José. Photograph by Richard Masoner, courtesy of CC 2.0.

Trees sequester carbon, provide shade, capture air pollutants, provide habitat, and offer various other benefits to urban and rural communities.

Including more locally native tree species in city-approved street tree palettes can help maximize ecosystem services while benefiting native wildlife. Larger trees confer greater benefits, so **providing trees with adequate growing conditions and conserving them as they age can bolster the benefits of the urban forest.** Municipal tree ordinances can be useful tools for curbing losses, in addition to setting **high tree replacement ratios** (i.e., the number of trees that must be planted for each existing tree removed) to expand the urban forest in private land and meet **city-scale canopy cover targets.** Moreover, tax credits and other incentive-based programs can be used to encourage people to plant and maintain trees on their private properties.

In San José especially, but also across California and the United States, the **large-scale redevelopment** of office parks is an opportunity to add housing close to job centers (thus reducing transportation burdens) and to build public and private partnerships for natural infrastructure.

Cities can leverage these development opportunities to build more affordable housing and invest in parks, stormwater management, and other nature-based solutions. New residential developments in San José already help fund park creation and improvements through **park dedicated fees**. Similar fees for commercial development can likewise support the local park system. Cities could also incentivize open space creation in private land through floor-area-ratio bonuses and transferable development programs. Making these new greenspaces publicly accessible would extend their benefits to the greater community.

Top: Santana Row, San José. Photograph by Eric Fredericks, courtesy of CC 2.0.
Bottom: Sunnyvale. Imagery courtesy of Google Earth.

Earth Day march, East Oakland. Photograph by Black Hour, courtesy of CC 2.0.

Local and state governments need to work proactively to combat **environmental injustice** and prioritize greening projects in underserved neighborhoods which are disproportionately affected by climate risk.

Without accompanying anti-displacement policies, greening in disadvantaged communities can lead to increased property values and gentrification (Rigolon and Németh 2018). Meaningful community engagement, affordable housing, and anti-displacement tools can help prevent gentrification and displacement. Planners and designers should partner with communities early in the process to gain local spatial knowledge and set objectives, as well as to demonstrate how stakeholder feedback will be meaningfully incorporated. Moreover, greening efforts and master plans will be most effective at reducing historical disparities in environmental services when developed **in conjunction with equity plans for housing** (e.g., rental subsidies, rent control, and community land trusts) **and job** (e.g., job training and creation, support for small businesses, and local hiring).

Street garden circle
provides bioretention
services, Berkeley.
Photograph by Robin
Grossinger, SFEI.

Aligning policies across levels and agencies of government may increase the success of planning efforts to build community resilience with nature.

Through general plans, city and county governments can establish a foundational framework that guides subordinate building codes, zoning, and land-use guidelines to better support nature-based solutions, increase density, and protect open space. **General plans can promote compact development, protect agricultural land and natural areas, and set minimum and maximum density targets to focus growth strategically around transportation nodes.** Changes in single-use zoning are another way that San José can catalyze densification and mixing of land uses. Minneapolis and other cities in the United States have demonstrated that this is a promising planning tool to encourage the diversification of housing stock, reduce commute times, and integrate supporting amenities (e.g., retail, office, and commercial development) near housing. However, realizing the shift away from suburban living would require substantial attitudinal changes among San José's current and future residents and significant community outreach. Highlighting the climate and health benefits of denser living, potential reduction in homeowner association fees, and diversification of housing stock that allows seniors to age in place would play a crucial role in this process.

Overlay zoning can add planning controls without changing land-use designation to implement specific resilience objectives and protect sensitive habitat.

In the case of Coyote Valley, the County's General Plan could establish a climate resilience, food security, and safety overlay to ensure working and natural lands are protected and preserved for climate change resilience, agriculture, flood control, groundwater protection, wildlife movement, habitat, and recreation. The overlay could establish performance standards to protect key environmental features and make room for natural processes, restrict and discourage development and intensive land uses, and attract funding from a climate resilience bond that could support such overlay zoning.

Many of these strategies require buy-in from multiple landowners.

To enact them, public agencies need to provide compelling **incentives for landowners** to adopt climate-smart and ecosystem-based management practices and need to educate the public about the value of nature-based solutions. Local and state governments can support these measures financially via **bond measures, mitigation credit programs, transferable development rights, cap and trade markets, and payment of ecosystem services programs.**

Public agencies and private actors have limited resources for greening interventions, which is why **interdisciplinary collaboration and landscape-scale planning** are key to ensure benefits are optimized across sectors.

This report outlines holistic strategies based on the collaboration of a science nonprofit, urban planning think tank, and open space agency, as well as the technical input of regulators, planners, scientists, and community groups. Scientists can develop quantifiable metrics and outcomes to help optimize benefits at the landscape scale and inform decisions made by separate authorities; planners and regulators can advance policy pathways to implement innovative nature-based solutions and promote greater alignment across government sectors; and trusts and open space agencies can help protect natural resources, connect people with nature, foster environmental stewardship, and work directly with landowners to implement resilience projects. By working together, these traditionally siloed sectors can develop strategy and policy documents that provide integrated technical guidance and reflect community priorities.

Residential vegetable garden in Oakland. Photograph by Boulosa, courtesy of CC 2.0.

Policy and Planning Recommendations	Place-types				
	UN	OP	CdS	CL	RO
Recommendation 1: Promote community-driven planning with robust engagement processes. Consult stakeholder groups early in the process to set goals and priorities, as well as to gather local spatial knowledge. Explore opportunities for community-based nonprofits to identify and organize neighborhood actions. Develop long-term public education programs around nature-based solutions and ecosystem services to catalyze their implementation.	X	X	X	X	X
Recommendation 2: Develop housing and job equity plans alongside greening projects, so that existing communities can benefit from greening interventions.	X	X	X	X	X
Recommendation 3: Eliminate parking minimums to promote walkability, free up space for additional development and urban greening features, and reduce the area of surface parking lots.	X	X	X		
Recommendation 4: Leverage new development to fund publicly accessible greenspace creation and improvement by offering incentives (e.g., floor-area-ratio bonuses) and commercial park fee requirements. Introduce green area ratio (percent of lot dedicated to greenspace) regulations for development permit applications.	X	X	X		
Recommendation 5: Set high tree replacement ratios to preserve and expand the urban forest. Increase requirements for tree basin size that promote tree health and longer lifespans. Promote the use of native and site-appropriate tree species with broad canopies to better support local biodiversity, reduce heat stress, and improve water efficiency.	X	X	X		
Recommendation 6: Establish park, streetscape, and stormwater design guidelines that prioritize local native species, mimic native habitat structure, buffer wetlands, and promote wildlife-friendly management practices.	X	X	X		
Recommendation 7: Develop county, city, and neighborhood master plans that can guide new development by identifying cool islands and hot spots, major sources of air pollution, areas prone to flooding, ecological corridors, habitat zones, and priority areas for park creation that deliver greenspaces within a ten minute walk of all residents.	X	X	X	X	X
Recommendation 8: Master plans should set a target of establishing at least half of new green features in communities designated as disadvantaged communities by CalEnviroScreen 3.0, or more locally relevant definitions of disadvantaged communities, such as household income below 80% of the County median.	X	X	X	X	X
Recommendation 9: Create policies to move utilities underground for any redevelopment, including high-end development projects and existing residential neighborhoods. This also helps to reduce fire risk.	X	X	X	X	X
Recommendation 10: Set targets to increase overall tree canopy cover to 40% to mitigate the urban heat island effect, reduce air pollution, capture runoff, and sequester carbon. Develop ordinances that protect existing large urban trees that provide ecological value to wildlife.	X	X	X		
Recommendation 11: Develop climate adaptation plans that integrate biodiversity and ecosystem service planning.	X	X	X	X	X
Recommendation 12: Create habitat protection overlay zones along sensitive habitat, such as wetlands and stream corridors, to limit building encroachment, light pollution, and other environmental stressors.	X	X	X	X	X
Recommendation 13: Enhance the ecological value and reduce the water consumption of existing greenspaces by implementing native habitat restoration strategies, such as re-oaking. Build 'seed' parks in neighborhoods to pilot initiatives and garner community support.	X	X	X		

P	Priority (*** = most important)	Ease of Implementation (*** = easiest, * = hardest)	Key Parties			
			City of San José	Santa Clara County	Individual (Resident, Business Owner, Developer etc.)	Other Partners (community based organizations, nonprofits etc.)
X	•••	•	X	X	X	X
	•••	••	X	X		X
	••	••	X			
	••	•	X	X		
	••	•	X	X		
	••	••	X			X
X	•••	••	X	X	X	X
X	••	••	X	X		X
X	•	•	X	X		
	•••	••	X			
X	•••	••	X	X		X
X	•••	••	X	X		X
X	••	•	X	X	X	X

Policy and Planning Recommendations	Place-types				
	UN	OP	CdS	CL	RO
Recommendation 14: Encourage policies that modify existing zoning to facilitate a mixture of land uses (e.g., multi-family units, retail, commercial, and office space) and reduce commute times. Promote infill development within existing building footprints, when possible and appropriate, to avoid greenspace loss.	X	X	X		
Recommendation 15: Develop programs to buy back private land at market rates for ecosystem function, such as in floodplains, to establish publicly accessible greenspaces that provide recreational opportunities and parcels at the end of cul-de-sacs for gateways to trail networks that improve pedestrian circulation, and to facilitate wildlife movement.	X	X	X	X	X
Recommendation 16: Offer incentives for homeowners to increase permeable cover, plant and protect trees, select native species, and maintain greenspace in their private property. Organize native plant giveaway events that make it easier for people to access site-appropriate plant palettes. Up-front rebates and free giveaway programs should be prioritized to ensure that low-income families can access these programs. Create additonal incentives that are appealing to rental property owners.	X	X	X	X	X
Recommendation 17: Create general plan overlays for climate resiliency, agricultural land preservation, and/or habitat connectivity based on ecosystem service valuation. Consider setbacks and buffers from rural residential, agricultural facilities, and greenhouses to increase landscape connectivity for wildlife.			X	X	X
Recommendation 18: Create a Transfer of Development Rights or Carbon Credit program that provides landowners payment for extinguishment of development rights in rural areas, and encourages investment in enhanced land management practices.	X	X		X	X
Recommendation 19: Develop a Payment for Ecosystem Services program, perhaps through an Agricultural Enterprise Zone, that recognizes natural infrastructure benefits provided by responsible / regenerative agricultural practices, such as sequestration of soil carbon, planting of wildlife-friendly crops, and groundwater recharge during flood events.				X	X
Recommendation 20: Reduce speed limits to reduce and slow traffic along roads to allow for movement of slower-moving agricultural equipment and reduce road mortality for wildlife. Build wildlife crossing infrastructure outside of roadways to create wildlife corridors that link populations in the Diablo Range and Santa Cruz Mountains.				X	X
Recommendation 21: Fund adaptation studies to help landowners understand best management practices. Develop an agricultural incubator to test innovative land management practices and quantify benefits.				X	X
Recommendation 22: Develop agricultural conservation easements and other voluntary financial incentives for landowners to protect open space and enhance ecological connectivity. Offer funding to help farmers transition to more climate-appropriate crops.				X	X
Recommendation 23: Coordinate with any regional agriculture planning efforts to emphasize need for comprehensive agricultural sustainability planning that includes preservation of prime farmland, appropriate use of less productive lands for processing and other infrastructure facilities, and local farmworker housing.				X	X
Recommendation 24: Develop a master plan vision for the phased restoration of newly protected land with the help of an interdisciplinary group of experts.					
Recommendation 25: Build ecological education centers to provide opportunities for people to learn first-hand the value of natural infrastructure and promote environmental stewardship.	X			X	X

P	Priority (*** = most important)	Ease of Implementation (*** = easiest, * = hardest)	City of San José	Santa Clara County	Individual (Resident, Business Owner, Developer etc.)	Other Partners (community based organizations, nonprofits etc.)
	•••	•	X	X		X
	•••	•••	X			X
	••	•••	X	X		X
X	•••	••	X			X
X	••	•	X	X		X
X	••	•	X	X		X
X	••	•••	X	X		
X	••	••		X	X	X
P	•••	••	X	X	X	X
X	•••	•		X		X
X	•••	••	X	X		X
X	••	•••	X	X		X

5 CONCLUSION

Climate change poses an imminent threat to urban and rural communities alike. To date, climate adaptation has largely focused on small-scale, engineered solutions to mitigate local risks. However, using nature-based solutions rather than gray infrastructure and coordinating actions across the urban-rural divide can maximize communities' preparedness for future climate conditions. Enhancing the ecological value of cities and rural areas can simultaneously provide economic, social, and environmental benefits, from bolstering regional biodiversity and creating jobs to improving human health and well-being.

Increasing density does not have to come at the expense of natural capital. In the Urban Neighborhood study cell, the number of residents and jobs doubled while greenspace area tripled. Tree canopy cover grew from 15 to 47%, which quadrupled carbon storage and would significantly help mitigate rising temperatures and extreme heat events. Impervious cover in the office park cell was transformed into 37 new acres of greenspace for residents and workers, delivering more than five times the greenspace per capita target set by the state of California and creating a local biodiversity and ecosystem service hub. A combination of tree plantings in private and public land in the Cul-de-Sac Suburb cell resulted in more than double the carbon sequestration, carbon storage, and avoided runoff, and nearly triple the amount of air pollutants removed annually. When aggregated, these interventions can help build regional climate resilience, especially when coordinated with improvements in the urban periphery.

Open space and agriculture protection is crucial to ensure the preservation of vital ecosystem services such as carbon sequestration and storage, local food production, flood-risk-management, and habitat provision. Incorporating nature-based solutions in rural areas can help amplify ecosystem-service delivery. In the Parks & Protected Areas study cell, the restoration of wetlands and other historical habitat types has the potential to protect critical wildlife linkages while functioning as regional stormwater infrastructure that benefits urban areas downstream. The adoption of climate-friendly practices in the Rural & Open Space and Cultivated Land study cells, could also help build resilience and reduce wildlife conflict. Collectively, landowners could get up to two million dollars in carbon payments annually for establishing hedgerows, mulching, applying compost, and building riparian buffers.

The strategies and solutions presented in this report vary widely in their implementation costs and associated benefits, though **larger-scale actions generally reduce risks from floods, heat waves, droughts, and fires more effectively than more localized ones.** Various policy tools exist to support these large-scale actions, but realizing them will nonetheless require concerted efforts from multiple levels of government and buy-in from diverse stakeholder groups.

Left top: Bobcat in Santa Teresa County Park above Coyote Creek Valley. Photograph by Don DeBold, courtesy of CC 2.0. Left bottom: Top Leaf Farm, rooftop in San Francisco. Photograph by Eli Zigas, courtesy of SPUR.

6 REFERENCES

[ABAG] Association of Bay Area Governments. 2006. "Planned Land Use." San Francisco, CA: Association of Bay Area Governments. http://opendata.mtc.ca.gov/datasets/planned-land-use-2006.

Abercrombie, Lauren C., James F. Sallis, Terry L. Conway, Lawrence D. Frank, Brian E. Saelens, and James E. Chapman. 2008. "Income and Racial Disparities in Access to Public Parks and Private Recreation Facilities." *American Journal of Preventive Medicine* 34 (1): 9–15. https://doi.org/10.1016/j.amepre.2007.09.030.

Akbari, Hashem, and Dionysia Kolokotsa. 2016. "Three Decades of Urban Heat Islands and Mitigation Technologies Research." *Energy and Buildings* 133 (December): 834–42. https://doi.org/10.1016/j.enbuild.2016.09.067.

Arslan, Selçuk, and Ali Aybek. 2012. "Particulate Matter Exposure in Agriculture." *Air Pollution - A Comprehensive Perspective*, August. https://doi.org/10.5772/50084.

Baruchman, Michelle. 2020. "Seattle Will Permanently Close 20 Miles of Residential Streets to Most Vehicle Traffic." *The Seattle Times*, May 7, 2020, sec. Traffic Lab. https://www.seattletimes.com/seattle-news/transportation/seattle-will-permanently-close-20-miles-of-residential-streets-to-most-vehicle-traffic/.

Basu, Rupa, Wen-Ying Feng, and Bart D. Ostro. 2008. "Characterizing Temperature and Mortality in Nine California Counties." *Epidemiology* 19 (1): 138–45.

Bay Area Open Space Council. 2019. "The Conservation Lands Network 2.0 Report: A Regional Conservation Strategy for the San Francisco Bay Area." Berkeley, CA: Bay Area Open Space Council.

Beagle, Julie, Jeremy Lowe, Katie McKnight, Sam Safran, Laura Tam, and Sara Jo Szambelan. 2019. "San Francisco Bay Shoreline Adaptation Atlas: Working with Nature to Plan for Sea Level Rise Using Operational Landscape Units." 915. Richmond, CA: San Francisco Estuary Institute.

Beil, Kurt, and Douglas Hanes. 2013. "The Influence of Urban Natural and Built Environments on Physiological and Psychological Measures of Stress— A Pilot Study." *International Journal of Environmental Research and Public Health* 10 (4): 1250–67. https://doi.org/10.3390/ijerph10041250.

Beller, Erin, Micha Salomon, and Robin Grossinger. 2010. "Historical Vegetation and Drainage Patterns of Western Santa Clara Valley: A Technical Memorandum Describing Landscape Ecology in Lower Peninsula, West Valley, and Guadalupe Watershed Management Areas." 622. Oakland, CA: SFEI.

Beller, Erin, Kelly, Maggi, and Larsen, Laurel G. 2020. "From savanna to suburb: Effects of 160 years of landscape change on carbon storage in Silicon Valley, California." Landscape and Urban Planning 195 (March): https://doi.org/10.1016/j.landurbplan.2019.103712.

Berg, Neil, and Alex Hall. 2015. "Increased Interannual Precipitation Extremes over California under Climate Change." *Journal of Climate* 28 (16): 6324–34. https://doi.org/10.1175/JCLI-D-14-00624.1.

Bettencourt, Luís M. A., José Lobo, Dirk Helbing, Christian Kühnert, and Geoffrey B. West. 2007. "Growth, Innovation, Scaling, and the Pace of Life in Cities." *Proceedings of the National Academy of Sciences* 104 (17): 7301–6. https://doi.org/10.1073/pnas.0610172104.

Bliss, Laura. 2020. "How Oakland Made Pedestrian-Friendly Slow Streets." *Bloomberg CityLab*, April 17, 2020. https://www.bloomberg.com/news/articles/2020-04-17/how-oakland-made-pedestrian-friendly-slow-streets.

Bowman, David M. J. S., Jennifer Balch, Paulo Artaxo, William J. Bond, Mark A. Cochrane, Carla M. D'Antonio, Ruth DeFries, et al. 2011. "The Human Dimension of Fire Regimes on Earth." *Journal of Biogeography* 38 (12): 2223–36. https://doi.org/10.1111/j.1365-2699.2011.02595.x.

Brans, Kristien I., Mieke Jansen, Joost Vanoverbeke, Nedim Tüzün, Robby Stoks, and Luc De Meester. 2017. "The Heat Is on: Genetic Adaptation to Urbanization Mediated by Thermal Tolerance and Body Size." *Global Change Biology* 23 (12): 5218–27. https://doi.org/10.1111/gcb.13784.

Burton, Elizabeth. 2002. "Measuring Urban Compactness in UK Towns and Cities." *Environment and Planning B: Planning and Design* 29 (2): 219–50. https://doi.org/10.1068/b2713.

[Cal-Adapt]. 2020. "Extreme Heat Days & Warm Nights." Berkeley, CA: UC Berkeley Geospatial Innovation Facility. https://cal-adapt.org/tools/extreme-heat/.

CalEPA. 2018. "California Climate Investments to Benefit Disadvantaged Communities." California Environmental Protection Agency. June 2018. https://calepa.ca.gov/envjustice/ghginvest/.

Calfapietra, C., S. Fares, F. Manes, A. Morani, G. Sgrigna, and F. Loreto. 2013. "Role of Biogenic Volatile Organic Compounds (BVOC) Emitted by Urban Trees on Ozone Concentration in Cities: A Review." *Environmental Pollution (Barking, Essex: 1987)* 183 (December): 71–80. https://doi.org/10.1016/j.envpol.2013.03.012.

CalFire. 2007. "Maps of Fire Hazard Severity Zones in the Statewide Responsibility Area." 1:1,000,000. Sacramento, CA: CalFire. https://osfm.fire.ca.gov/divisions/wildfire-planning-engineering/wildland-hazards-building-codes/fire-hazard-severity-zones-maps/.

[CARB] California Air Resources Board. 2018. "Priority Population Investments." Sacramento, CA: California Air Resources Board. https://ww3.arb.ca.gov/cc/capandtrade/auctionproceeds/communityinvestments.htm.

Chaplin-Kramer, Rebecca, and Melvin R. George. 2013. "Effects of Climate Change on Range Forage Production in the San Francisco Bay Area." *PLoS ONE* 8 (3). https://doi.org/10.1371/journal.pone.0057723.

City of San José. 2015. "Street Pavement Maintenance: Road Condition Is Deteriorating Due to Insufficient Funding." 15–02. San José, CA: City of San José, Office of the City Auditor.

———. 2016. "Appendix D: Community-Wide GHG Emissions Inventory and Forecasts Memo." San José, CA: City of San José, California.

———. 2018. "Street Trees in San Jose, CA Maintained or Supervised by the Department of Transportation." San José, CA: City of San José, Department of Transportation.

Clausnitzer, H., and M. J. Singer. 1996. "Respirable-Dust Production from Agricultural Operations in the Sacramento Valley, California." *Journal of Environmental Quality* 25 (4): 877–84. https://doi.org/10.2134/jeq1996.00472425002500040032x.

Cohen, Jack D. 2000. "Preventing Disaster: Home Ignitability in the Wildland-Urban Interface." *Journal of Forestry* 98 (3): 15–21. https://doi.org/10.1093/jof/98.3.15.

Cohen-Shacham, E., G. Walters, C. Janzen, and S. Maginnis, eds. 2016. *Nature-Based Solutions to Address Global Societal Challenges*. IUCN International Union for Conservation of Nature. https://doi.org/10.2305/IUCN.CH.2016.13.en.

Cooley, Heather, Kristina Donnelly, Salote Soqo, and Colin Bailey. 2016. "Drought and Equity in the San Francisco Bay Area." 978-1-893790-73–5. Oakland, CA: Pacific Institute.

Coombs, James Scott, and John M. Melack. 2013. "Initial Impacts of a Wildfire on Hydrology and Suspended Sediment and Nutrient Export in California Chaparral Watersheds." *Hydrological Processes* 27 (26): 3842–51. https://doi.org/10.1002/hyp.9508.

Corburn, Jason. 2009. "Cities, Climate Change and Urban Heat Island Mitigation: Localising Global Environmental Science." *Urban Studies* 46 (2): 413–27. https://doi.org/10.1177/0042098008099361.

Costanza, Robert, Ralph d'Arge, Rudolf de Groot, Stephen Farber, Monica Grasso, Bruce Hannon, Karin Limburg, et al. 1997. "The Value of the World's Ecosystem Services and Natural Capital." *Nature* 387 (6630): 253–60. https://doi.org/10.1038/387253a0.

[CPAD] GreenInfo Network. 2020. "California Protected Areas Database." Oakland, CA: GreenInfo Network.

Cushing, Lara, Dan Blaustein-Rejto, Madeline Wander, Manuel Pastor, James Sadd, Allen Zhu, and Rachel Morello-Frosch. 2018. "Carbon Trading, Co-Pollutants, and Environmental Equity: Evidence from California's Cap-and-Trade Program (2011–2015)." *PLOS Medicine* 15 (7): e1002604. https://doi.org/10.1371/journal.pmed.1002604.

Davis, J. B. 1990. "The Wildland-Urban Interface: Paradise or Battleground?" *Journal of Forestry* 88 (1): 26–31.

Debbage, Neil, and J. Marshall Shepherd. 2015. "The Urban Heat Island Effect and City Contiguity." *Computers, Environment and Urban Systems* 54 (November): 181–94. https://doi.org/10.1016/j.compenvurbsys.2015.08.002.

DeJong, Ted M. 2005. "Using Physiological Concepts to Understand Early Spring Temperature Effects on Fruit Growth and Anticipating Fruit Size Problems at Harvest," 4.

Depietri, Yaella, and Timon McPhearson. 2017. "Integrating the Grey, Green, and Blue in Cities: Nature-Based Solutions for Climate Change Adaptation and Risk Reduction." In *Nature-Based Solutions to Climate Change Adaptation in Urban Areas: Linkages between Science, Policy and Practice*, edited by Nadja Kabisch, Horst Korn, Jutta Stadler, and Aletta Bonn, 91–109. Theory and Practice of Urban Sustainability Transitions. Cham: Springer International Publishing. https://doi.org/10.1007/978-3-319-56091-5_6.

Dettinger, Michael. 2011. "Climate Change, Atmospheric Rivers, and Floods in California – A Multimodel Analysis of Storm Frequency and Magnitude Changes1." *JAWRA Journal of the American Water Resources Association* 47 (3): 514–23. https://doi.org/10.1111/j.1752-1688.2011.00546.x.

Diffenbaugh, Noah S., Daniel L. Swain, and Danielle Touma. 2015. "Anthropogenic Warming Has Increased Drought Risk in California." *Proceedings of the National Academy of Sciences* 112 (13): 3931–36. https://doi.org/10.1073/pnas.1422385112.

Donovan, Geoffrey H., David T. Butry, Yvonne L. Michael, Jeffrey P. Prestemon, Andrew M. Liebhold, Demetrios Gatziolis, and Megan Y. Mao. 2013. "The Relationship Between Trees and Human Health: Evidence from the Spread of the Emerald Ash Borer." *American Journal of Preventive Medicine* 44 (2): 139–45. https://doi.org/10.1016/j.amepre.2012.09.066.

Dreier, E. P. Clapp Distinguished Professor of Politics Peter, Peter Dreier, John H. Mollenkopf, and Todd Swanstrom. 2004. *Place Matters: Metropolitics for the Twenty-First Century.* University Press of Kansas.

Dueñas, Norberto. 2016. "Comments on CalEnviroScreen 3.0." San José, CA: City of San José, Office of the City Manager.

Earman, Sam, and Michael Dettinger. 2011. "Potential Impacts of Climate Change on Groundwater Resources – a Global Review." *Journal of Water and Climate Change* 2 (4): 213–29. https://doi.org/10.2166/wcc.2011.034.

EarthDefine. 2013. "SpatialCover Tree Canopy." Redmond, WA: EarthDefine LLC. https://www.earthdefine.com/spatialcover_treecanopy/.

Ewing, Reid, Shima Hamidi, Guang Tian, David Proffitt, Stefania Tonin, and Laura Fregolent. 2018. "Testing Newman and Kenworthy's Theory of Density and Automobile Dependence." *Journal of Planning Education and Research* 38 (2): 167–82. https://doi.org/10.1177/0739456X16688767.

Ewing, Reid, Richard A. Schieber, and Charles V. Zegeer. 2003. "Urban Sprawl as a Risk Factor in Motor Vehicle Occupant and Pedestrian Fatalities." *American Journal of Public Health* 93 (9): 1541–45. https://doi.org/10.2105/AJPH.93.9.1541.

Ewing, Reid, Tom Schmid, Richard Killingsworth, Amy Zlot, and Stephen Raudenbush. 2003. "Relationship between Urban Sprawl and Physical Activity, Obesity, and Morbidity." *American Journal of Health Promotion* 18 (1): 47–57. https://doi.org/10.4278/0890-1171-18.1.47.

Fanai, Amir, Sukarn Claire, Tan Dinh, Michael Nguyen, and Stuart Schultz. 2014. "Bay Area Emissions Inventory Summary Report: Criteria Air Pollutants, Base Year 2011." San Francisco, CA: Bay Area Air Quality Management District.

Faust, John, Laura August, Komal Bangia, Vanessa Galaviz, Julian Leichty, Shankar Prasad, Rose Schmitz, et al. 2017. "CalEnviroScreen 3.0." Sacramento, CA: California Environmental Protection Agency and Office of Environmental Health Hazard Assessment.

FEMA. 2009. "Santa Clara County Flood Risk Map." Washington, D.C.: Federal Emergency Management Agency.

Fitzky, Anne Charlott, Hans Sandén, Thomas Karl, Silvano Fares, Carlo Calfapietra, Rüdiger Grote, Amélie Saunier, and Boris Rewald. 2019. "The Interplay Between Ozone and Urban Vegetation—BVOC Emissions, Ozone Deposition, and Tree Ecophysiology." *Frontiers in Forests and Global Change* 2. https://doi.org/10.3389/ffgc.2019.00050.

Flörke, Martina, Christof Schneider, and Robert I. McDonald. 2018. "Water Competition between Cities and Agriculture Driven by Climate Change and Urban Growth." *Nature Sustainability* 1 (1): 51–58. https://doi.org/10.1038/s41893-017-0006-8.

[FMMP] Farmland Mapping and Monitoring Program. 2018. "California Important Farmland." Sacramento, CA: California Department of Conservation, Division of Land Resource Protection, Farmland Mapping and Monitoring Program.

Freidin, Robert, Diane Schreck, Brooke Scruggs, Elise Shulman, Alissa Swauger, and Allison Tashnek. 2011. "Wildlife Use of the Los Piñetos Underpass Santa Clarita, California." Los Angeles, CA: UCLA Institute of the Environment and Sustainability.

Fried, Jeremy S., Margaret S. Torn, and Evan Mills. 2004. "The Impact of Climate Change on Wildfire Severity: A Regional Forecast for Northern California." *Climatic Change* 64 (1): 169–91. https://doi.org/10.1023/B:CLIM.0000024667.89579.ed.

Frumkin, Howard. 2002. "Urban Sprawl and Public Health." *Public Health Reports* 117 (3): 201–17. https://doi.org/10.1093/phr/117.3.201.

Furman Center. 2020. "COVID-19 Cases in New York City, a Neighborhood-Level Analysis." *The Stoop: NYU Furman Center Blog* (blog). April 10, 2020. https://furmancenter.org/thestoop/entry/covid-19-cases-in-new-york-city-a-neighborhood-level-analysis.

Gago, E. J., J. Roldan, R. Pacheco-Torres, and J. Ordóñez. 2013. "The City and Urban Heat Islands: A Review of Strategies to Mitigate Adverse Effects." *Renewable and Sustainable Energy Reviews* 25 (September): 749–58. https://doi.org/10.1016/j.rser.2013.05.057.

Girard, Kirk, Rob Eastwood, Manira Sandhir, Charu Ahluwalia, Steve Borgstrom, Joe Deviney, Andrea Mackenzie, Matt Freeman, and Jake Smith. 2018. "Santa Clara Valley Agricultural Plan: Investing in Our Working Lands for Regional Resilience." San José, CA: Santa Clara Valley Open Space Authority.

Grant, B., Sara Jo Szambelan, Stephen Engblom, Cristian Bevington, and Hugo Errazuriz. 2020. "Model Places: Envisioning a future Bay Area with room and opportunity for everyone." San Francisco, CA: SPUR.

Gravuer, Kelly. 2016. "Compost Application Rates for California Croplands and Rangelands for a CDFA Healthy Soils Incentives Program." Sacramento, CA: California Department of Food and Agriculture.

Griscom, Bronson W., Justin Adams, Peter W. Ellis, Richard A. Houghton, Guy Lomax, Daniela A. Miteva, William H. Schlesinger, et al. 2017. "Natural Climate Solutions." *Proceedings of the National Academy of Sciences* 114 (44): 11645–50. https://doi.org/10.1073/pnas.1710465114.

Grossinger, Robin, Ruth Askevold, Chuck Striplen, Elise Brewster, Sarah Pearce, Kristen Larned, Lester McKee, and Josh Collins. 2006. "Coyote Creek Watershed Historical Ecology Study: Historical Condition, Landscape Change, and Restoration Potential in the Eastern Santa Clara Valley, California." 426. Oakland, CA: San Francisco Estuary Institute.

Guirguis, Kristen, Alexander Gershunov, Alexander Tardy, and Rupa Basu. 2014. "The Impact of Recent Heat Waves on Human Health in California." *Journal of Applied Meteorology and Climatology* 53 (1): 3–19. https://doi.org/10.1175/JAMC-D-13-0130.1.

Haaland, Christine, and Cecil Konijnendijk van den Bosch. 2015. "Challenges and Strategies for Urban Green-Space Planning in Cities Undergoing Densification: A Review." *Urban Forestry & Urban Greening* 14 (4): 760–71. https://doi.org/10.1016/j.ufug.2015.07.009.

Hajat, Shakoor, and Tom Kosatky. 2009. "Heat-Related Mortality: A Review and Exploration of Heterogeneity." *Journal of Epidemiology and Community Health* 64 (September): 753–60. https://doi.org/10.1136/jech.2009.087999.

Harlan, Sharon L, and Darren M Ruddell. 2011. "Climate Change and Health in Cities: Impacts of Heat and Air Pollution and Potential Co-Benefits from Mitigation and Adaptation." *Current Opinion in Environmental Sustainability* 3 (3): 126–34. https://doi.org/10.1016/j.cosust.2011.01.001.

He, Minxue, and Mahesh Gautam. 2016. "Variability and Trends in Precipitation, Temperature and Drought Indices in the State of California." *Hydrology* 3 (2): 14. https://doi.org/10.3390/hydrology3020014.

Howitt, Richard, Duncan MacEwan, Josué Medellín-Azuara, Jay Lund, and Daniel Sumner. 2015. "Economic Analysis of the 2015 Drought For California Agriculture." Davis, CA: UC Davis Center for Watershed Sciences.

ICF International. 2012. "Chapter 3: Physical and Biological Resources." In *Santa Clara Valley Habitat Plan*. San Francisco, CA: ICF International.

Jackson, L. E., S. M. Wheeler, A. D. Hollander, A. T. O'Geen, B. S. Orlove, J. Six, D. A. Sumner, et al. 2011. "Case Study on Potential Agricultural Responses to Climate Change in a California Landscape." *Climatic Change* 109 (1): 407–27. https://doi.org/10.1007/s10584-011-0306-3.

Jacob, Daniel J., and Darrell A. Winner. 2009. "Effect of Climate Change on Air Quality." *Atmospheric Environment*, Atmospheric Environment - Fifty Years of Endeavour, 43 (1): 51–63. https://doi.org/10.1016/j.atmosenv.2008.09.051.

Kassab, Bassam, George Cook, Chanie Abuye, Benjamin Apolo, Henry Barrientos, Victoria García, Ardy Ghoreishi, Simon Gutierrez, and Tracy Hemmeter. 2016. "2016 Groundwater Management Plan: Santa Clara and Llagas Subbasins." San José, CA: Santa Clara Valley Water District.

Keeler, Bonnie L., Perrine Hamel, Timon McPhearson, Maike H. Hamann, Marie L. Donahue, Kelly A. Meza Prado, Katie K. Arkema, et al. 2019. "Social-Ecological and Technological Factors Moderate the Value of Urban Nature." *Nature Sustainability* 2 (1): 29–38. https://doi.org/10.1038/s41893-018-0202-1.

Keeley, Jon E. 2003. "Fire and Invasive Plants in California Ecosystems." *Fire Management Today* 63 (2): 18–19.

Kheirbek, Iyad, Katherine Wheeler, Sarah Walters, Daniel Kass, and Thomas Matte. 2013. "PM2.5 and Ozone Health Impacts and Disparities in New York City: Sensitivity to Spatial and Temporal Resolution." *Air Quality, Atmosphere & Health* 6 (2): 473–86. https://doi.org/10.1007/s11869-012-0185-4.

Kondo, Michelle C., SeungHoon Han, Geoffrey H. Donovan, and John M. MacDonald. 2017. "The Association between Urban Trees and Crime: Evidence from the Spread of the Emerald Ash Borer in Cincinnati." *Landscape and Urban Planning* 157 (January): 193–99. https://doi.org/10.1016/j.landurbplan.2016.07.003.

Kravchenko, Julia, Amy P. Abernethy, Maria Fawzy, and H. Kim Lyerly. 2013. "Minimization of Heatwave Morbidity and Mortality." *American Journal of Preventive Medicine* 44 (3): 274–82. https://doi.org/10.1016/j.amepre.2012.11.015.

Krawchuk, Meg, and Max Moritz. 2012. "Fire and Climate Change in California: Changes in the Distribution and Frequency of Fire in Climates of the Future and Recent Past (1911–2099)." CEC-500-2012-026. Sacramento, CA: California Energy Commission.

Lachman, Steven Frederic. 2001. "Should Municpalities Be Liable for Development-Related Flooding." NATURAL RESOURCES JOURNAL 41: 37.

Laurenson, Georgina, Seth Laurenson, Nanthi Bolan, Simon Beecham, and Ian Clark. 2013. "Chapter Four - The Role of Bioretention Systems in the Treatment of Stormwater." In Advances in Agronomy, edited by Donald L. Sparks, 120:223–74. Academic Press. https://doi.org/10.1016/B978-0-12-407686-0.00004-X.

Lopez, Nadia. 2019. "Is Banning Single-Family Zoning Possible in San Jose?" San José Spotlight, July 22, 2019. https://sanjosespotlight.com/is-banning-single-family-zoning-possible-in-san-jose/.

Lund, Jay, Josue Medellin-Azuara, John Durand, and Kathleen Stone. 2018. "Lessons from California's 2012–2016 Drought." Journal of Water Resources Planning and Management 144 (10): 04018067. https://doi.org/10.1061/(ASCE)WR.1943-5452.0000984.

Mann, Michael L., Enric Batllori, Max A. Moritz, Eric K. Waller, Peter Berck, Alan L. Flint, Lorraine E. Flint, and Emmalee Dolfi. 2016. "Incorporating Anthropogenic Influences into Fire Probability Models: Effects of Human Activity and Climate Change on Fire Activity in California." PLOS ONE 11 (4): e0153589. https://doi.org/10.1371/journal.pone.0153589.

McDonald, Rob, and Erica Spotswood. 2020. "Cities Are Not to Blame for the Spread of COVID-19—nor Is the Demise of Cities an Appropriate Response." The Nature of Cities (blog). April 14, 2020. https://www.thenatureofcities.com/2020/04/14/cities-are-not-to-blame-for-the-spread-of-covid-19-nor-is-the-demise-of-cities-an-appropriate-response/.

McDonald, Robert I., Andressa V. Mansur, Fernando Ascensão, M'lisa Colbert, Katie Crossman, Thomas Elmqvist, Andrew Gonzalez, et al. 2020. "Research Gaps in Knowledge of the Impact of Urban Growth on Biodiversity." Nature Sustainability 3 (1): 16–24. https://doi.org/10.1038/s41893-019-0436-6.

McPherson, E. Gregory, Natalie S. van Doorn, and Paula J. Peper. 2016. "Urban Tree Database and Allometric Equations." General Technical Report PSW-GTR-253. Albany, CA: USDA Forest Service. https://www.fs.fed.us/psw/publications/documents/psw_gtr253/psw_gtr_253.pdf.

Microsoft. 2018. "US Building Footprints." Redmond, WA: Microsoft, Bing Maps.

Miller, Jeremy. 2019. "Major Land Deal for Conservation Announced in Silicon Valley." Bay Nature, November 5, 2019. https://baynature.org/article/the-last-big-save/.

Moody, John A., and Deborah A. Martin. 2009. "Synthesis of Sediment Yields after Wildland Fire in Different Rainfall Regimes in the Western United States." International Journal of Wildland Fire 18 (1): 96–115. https://doi.org/10.1071/WF07162.

Mount, Jeffrey, Brian Gray, Caitrin Chappelle, Greg Gartrell, Ted Grantham, Peter Moyle, Nathaniel E. Seavy, Leon Szeptycki, and Barton "Buzz" Thompson. 2017. "Managing California's Freshwater Ecosystems: Lessons from the 2012–16 Drought." Public Policy Institute of California.

Moyce, Sally, Diane Mitchell, Tracey Armitage, Daniel Tancredi, Jill Joseph, and Marc Schenker. 2017. "Heat Strain, Volume Depletion and Kidney Function in California Agricultural Workers." *Occupational and Environmental Medicine* 74 (6): 402–9. https://doi.org/10.1136/oemed-2016-103848.

[NLCD] National Land Cover Dataset. 2016. "Urban Imperviousness." Sioux Falls, SD: USGS Multi-Resolution Land Characteristics Consortium.

Norman, Jonathan, Heather L. MacLean, and Christopher A. Kennedy. 2006. "Comparing High and Low Residential Density: Life-Cycle Analysis of Energy Use and Greenhouse Gas Emissions." *Journal of Urban Planning and Development* 132 (1): 10–21. https://doi.org/10.1061/(ASCE)0733-9488(2006)132:1(10).

NRCS. 2011. "Cover Crop." 340. Conservation Practice Standard. Natural Resources Conservation Service.

———. 2012. "Hedgerow Planting." 422. Conservation Practice Standard. Natural Resources Conservation Service.

———. 2014. "Residue and Tillage Management, Reduced Till." 345. Conservation Practice Standard. Natural Resources Conservation Service.

———. 2017. "Prescribed Grazing." 528. Conservation Practice Standard. Natural Resources Conservation Service.

Oke, T. R. 1973. "City Size and the Urban Heat Island." *Atmospheric Environment (1967)* 7 (8): 769–79. https://doi.org/10.1016/0004-6981(73)90140-6.

———. 1982. "The Energetic Basis of the Urban Heat Island." *Quarterly Journal of the Royal Meteorological Society* 108 (455): 1–24. https://doi.org/10.1002/qj.49710845502.

Olson, C, M Landgraf, and J Perez. 2016. "Understanding Our Communities: A Community Assessment Project." Mountan View, CA: Santa Clara Valley Open Space Authority & Basecamp Strategies.

Ong, Boon Lay. 2003. "Green Plot Ratio: An Ecological Measure for Architecture and Urban Planning." *Landscape and Urban Planning* 63 (4): 197–211. https://doi.org/10.1016/S0169-2046(02)00191-3.

[OSA] Santa Clara Valley Open Space Authority. 2019. "Coyote Valley: A Case for Conservation." San José, CA: Santa Clara Valley Open Space Authority.

[OSA] Santa Clara Valley Open Space Authority, and Conservation Biology Institute. 2017. "Coyote Valley Landscape Linkage: A Vision for a Resilient, Multi-Benefit Landscape." San José, CA: Santa Clara Valley Open Space Authority.

Pathak, Tapan B., Mahesh L. Maskey, Jeffery A. Dahlberg, Faith Kearns, Khaled M. Bali, and Daniele Zaccaria. 2018. "Climate Change Trends and Impacts on California Agriculture: A Detailed Review." *Agronomy* 8 (3): 25. https://doi.org/10.3390/agronomy8030025.

Pecl, Gretta T., Miguel B. Araújo, Johann D. Bell, Julia Blanchard, Timothy C. Bonebrake, I-Ching Chen, Timothy D. Clark, et al. 2017. "Biodiversity Redistribution under Climate Change: Impacts on Ecosystems and Human Well-Being." *Science* 355 (6332): eaai9214. https://doi.org/10.1126/science.aai9214.

Popovich, Nadja, Blacki Migliozzi, Karthik Patanjali, Anjali Singhvi, and Jon Huang. 2019. "See How the World's Most Polluted Air Compares With Your City's." *The New York Times*, December 3, 2019, sec. Climate. https://www.nytimes.com/interactive/2019/12/02/climate/air-pollution-compare-ar-ul.html.

Pugh, Thomas A. M., A. Robert MacKenzie, J. Duncan Whyatt, and C. Nicholas Hewitt. 2012. "Effectiveness of Green Infrastructure for Improvement of Air Quality in Urban Street Canyons." *Environmental Science & Technology* 46 (14): 7692–99. https://doi.org/10.1021/es300826w.

Reid, Colleen E, Michael Brauer, Fay Johnston, Michael Jerrett, John R. Balmes, and Catherine Elliott. 2016. "Critical Review of Health Impacts of Wildfire Smoke Exposure." *Environmental Health Perspectives* 124 (9): 1334–43. https://doi.org/10.1289/ehp.1409277.

Rigolon, Alessandro, and Jeremy Németh. 2018. "'We're Not in the Business of Housing:' Environmental Gentrification and the Nonprofitization of Green Infrastructure Projects." *Cities* 81 (November): 71–80. https://doi.org/10.1016/j.cities.2018.03.016.

Rogers, Paul. 2019. "San Jose: Victims of 2017 Flood to Receive Payments." *Mercury News*, June 20, 2019. https://www.mercurynews.com/2019/06/20/san-jose-victims-of-2017-coyote-creek-flood-to-receive-payments/.

Roman, Lara A, John J. Battles, and Joe R. McBride. 2016. "Urban Tree Mortality: A Primer on Demographic Approaches." General Technical Report NRS-158. Philadelphia, PA: US Forest Service.

Romanow, Kerrie, Ashwini Kantak, Rosalynn Hughey, Kim Walesh, and Jim Ortbal. 2018. "Climate Smart San José: A People-Centered Plan for a Low-Carbon City." San José, CA: City of San José, California.

SAGE. 2012. "Coyote Valley Sustainable Agriculture & Conservation: Feasibility Study and Recommendations." Berkeley, CA: Sustainable Agricultural Education (SAGE).

Sailor, David J. 2011. "A Review of Methods for Estimating Anthropogenic Heat and Moisture Emissions in the Urban Environment." *International Journal of Climatology* 31 (2): 189–99. https://doi.org/10.1002/joc.2106.

Sakakibara, Yasushi. 1996. "A Numerical Study of the Effect of Urban Geometry upon the Surface Energy Budget." *Atmospheric Environment*, Conference on the Urban Thermal Environment Studies in Tohwa, 30 (3): 487–96. https://doi.org/10.1016/1352-2310(94)00150-2.

Salmond, Jennifer A., Marc Tadaki, Sotiris Vardoulakis, Katherine Arbuthnott, Andrew Coutts, Matthias Demuzere, Kim N. Dirks, et al. 2016. "Health and Climate Related Ecosystem Services Provided by Street Trees in the Urban Environment." *Environmental Health* 15 (1): S36. https://doi.org/10.1186/s12940-016-0103-6.

Sanderson, Eric W, Joseph Walston, and John G Robinson. 2018. "From Bottleneck to Breakthrough: Urbanization and the Future of Biodiversity Conservation." *Bioscience* 68 (6): 412–26. https://doi.org/10.1093/biosci/biy039.

Santa Clara County. 2011. "Health and Social Inequity in Santa Clara County." San José, CA: Santa Clara County Public Health Department.

———. 2016. "City and Small Area/Neighborhood Public Health Profiles." San José, CA: Santa Clara County Public Health Department. https://www.sccgov.org/sites/phd/hi/hd/Pages/city-profiles.aspx.

Schwarz, Kirsten, Michail Fragkias, Christopher G. Boone, Weiqi Zhou, Melissa McHale, J. Morgan Grove, Jarlath O'Neil-Dunne, et al. 2015. "Trees Grow on Money: Urban Tree Canopy Cover and Environmental Justice." *PLoS ONE* 10 (4). https://doi.org/10.1371/journal.pone.0122051.

Seewagen, Chad L., Christine D. Sheppard, Eric J. Slayton, and Christopher G. Guglielmo. 2011. "Plasma Metabolites and Mass Changes of Migratory Landbirds Indicate Adequate Stopover Refueling in a Heavily Urbanized Landscape." *The Condor* 113 (2): 284–97. https://doi.org/10.1525/cond.2011.100136.

Sharath, M, Rajendra Kumar, Rk Naresh, and Mandapelli Chandra. 2019. "Chapter – 7: Conservation Agriculture and Mulching Improving Soil Carbon Sequestration in Arable Cropping Systems. Chapter – 8: Organic, Inorganic Systems and Conservation Agriculture Enhancing Carbon Sequestration. Chapter – 9: Agro-Biodiversity in Sustainable Food and Nutrition Security: Towards a New Paradigm of Conservation Agriculture." In .

Shonkoff, Seth B., Rachel Morello-Frosch, Manuel Pastor, and James Sadd. 2011. "The Climate Gap: Environmental Health and Equity Implications of Climate Change and Mitigation Policies in California—a Review of the Literature." *Climatic Change* 109 (1): 485–503. https://doi.org/10.1007/s10584-011-0310-7.

Silvera Seamans, Georgia. 2013. "Mainstreaming the Environmental Benefits of Street Trees." *Urban Forestry & Urban Greening* 12 (1): 2–11. https://doi.org/10.1016/j.ufug.2012.08.004.

Sonneveld, Ben G.J.S, Max D Merbis, Amani Alfarra, İ. H. Olcay Ünver, Maria Felicia Arnal, and Food and Agriculture Organization of the United Nations. 2018. "Nature-Based Solutions for Agricultural Water Management and Food Security." FAO Land and Water Discussion Paper 12. Rome, Italy: Food and Agriculture Organization of the United Nations.

Spotswood, Erica, Robin Grossinger, Steve Hagerty, Erin Beller, April Robinson, and Letitia Grenier. 2017. "Re-Oaking Silicon Valley: Building Vibrant Cities with Nature." 825. Richmond, CA: San Francisco Estuary Institute.

Stienstra, Tom. 2020. "East Bay Parks May Be next to Close as Huge Crowds Create Health, Safety Issues." *San Francisco Chronicle*, March 22, 2020. https://www.sfchronicle.com/bayarea/article/East-Bay-parks-may-be-next-to-close-as-huge-15149356.php.

Stoecklin-Marois, Maria, Tamara Hennessy-Burt, Diane Mitchell, and Marc Schenker. 2013. "Heat-Related Illness Knowledge and Practices among California Hired Farm Workers in The MICASA Study." *Industrial Health* 51 (1): 47–55. https://doi.org/10.2486/indhealth.2012-0128.

Stone, Brian. 2008. "Urban Sprawl and Air Quality in Large US Cities." *Journal of Environmental Management* 86 (4): 688–98. https://doi.org/10.1016/j.jenvman.2006.12.034.

Stone, Brian, and Michael O. Rodgers. 2001. "Urban Form and Thermal Efficiency: How the Design of Cities Influences the Urban Heat Island Effect." *American Planning Association. Journal of the American Planning Association; Chicago* 67 (2): 186–98. http://dx.doi.org.libproxy.berkeley.edu/10.1080/01944360108976228.

Storrow, Benjamin. 2020. "Why CO2 Isn't Falling More during a Global Lockdown." *Scientific American*, April 24, 2020. https://www.scientificamerican.com/article/why-co2-isnt-falling-more-during-a-global-lockdown/.

Stott, Iain, Masashi Soga, Richard Inger, and Kevin J. Gaston. 2015. "Land Sparing Is Crucial for Urban Ecosystem Services." *Frontiers in Ecology and the Environment* 13 (7): 387–93. https://doi.org/10.1890/140286.

Sturm, R., and D. A. Cohen. 2004. "Suburban Sprawl and Physical and Mental Health." *Public Health* 118 (7): 488–96. https://doi.org/10.1016/j.puhe.2004.02.007.

Takano, T., K. Nakamura, and M. Watanabe. 2002. "Urban Residential Environments and Senior Citizens' Longevity in Megacity Areas: The Importance of Walkable Green Spaces." *Journal of Epidemiology & Community Health* 56 (12): 913–18. https://doi.org/10.1136/jech.56.12.913.

Tarnay, Leland. 2018. "Living with Fire (and Smoke) in California." Vallejo, CA: US Forest Service, Region 5 Remote Sensing Lab.

Thurlow, Noelle. 2019. "Why Saving Coyote Valley in San Jose Is So Important." *Peninsula Open Space Trust* (blog). July 23, 2019. https://openspacetrust.org/blog/saving-coyote-valley/.

Tratalos, Jamie, Richard A. Fuller, Philip H. Warren, Richard G. Davies, and Kevin J. Gaston. 2007. "Urban Form, Biodiversity Potential and Ecosystem Services." *Landscape and Urban Planning* 83 (4): 308–17. https://doi.org/10.1016/j.landurbplan.2007.05.003.

US Census Bureau. 2017. "2013-2017 American Community Survey 5-Year Estimates." Suitland, MD: US Census Bureau.

———. 2018. "2014-2018 American Community Survey 5-Year Estimates." Suitland, MD: US Census Bureau.

USGS. 2014. "San Francisco Bay Area Climate-Smart Watershed Analyst - Beta Release." Petaluma, CA: US Geological Survey. https://geo3.pointblue.org/watershed-analyst/index.php?polygon=tbc3&basin=2205300800&climatevar=ppt&future=GFDL_A2&hflen=10&hstart=1951&fstart=2070&flen=30&hlen=30.

Vaidyanathan, Ambarish, and Ambarish Vaidyanathan. 2013. "Evaluating Extreme Heat Event Definitions: Region-Specific Investigation of Extreme Heat and Heat-Related Mortality." In . AMS. https://ams.confex.com/ams/93Annual/webprogram/Paper215809.html.

Weaver, C. P., X.-Z. Liang, J. Zhu, P. J. Adams, P. Amar, J. Avise, M. Caughey, et al. 2009. "A Preliminary Synthesis of Modeled Climate Change Impacts on U.S. Regional Ozone Concentrations." *Bulletin of the American Meteorological Society* 90 (12): 1843–64. https://doi.org/10.1175/2009BAMS2568.1.

Westerling, A. L., and B. P. Bryant. 2008. "Climate Change and Wildfire in California." *Climatic Change* 87 (S1): 231–49. https://doi.org/10.1007/s10584-007-9363-z.

Willingham, AJ. 2018. "Smoke from the California Wildfires Is Visible in New York City." *CNN*, November 20, 2018. https://www.cnn.com/2018/11/20/us/california-wildfires-new-york-city-trnd/index.html.

Ziter, Carly D., Eric J. Pedersen, Christopher J. Kucharik, and Monica G. Turner. 2019. "Scale-Dependent Interactions between Tree Canopy Cover and Impervious Surfaces Reduce Daytime Urban Heat during Summer." *Proceedings of the National Academy of Sciences* 116 (15): 7575–80. https://doi.org/10.1073/pnas.1817561116.

7APPENDICES

Appendix 1: Ecosystem Service Quantification Methods

Urban place-types

CURRENT LANDSCAPE CONDITIONS

Urban Job Centers and Job Centers place-type cells in San José were used as a proxy for Urban Neighborhoods because the former were not included in SPUR's Model Places analysis. In the next fifty years, cells in Downtown San José also have the potential of transforming from Urban Job Centers and Job Centers to Urban Neighborhoods classification. Current urban greening conditions in all San José urban place-type cells (Urban Job Centers, Job Centers, Office Parks, and Cul-de-Sac Suburbs) were quantified using the California Protected Areas Database (2020) and EarthDefine SpatialCover Tree Datasets (2013). The California Protected Areas Database provided information on the current acreage of public open spaces, while EarthDefine SpatialCover Tree Datasets provided data on the number of trees and canopy cover. The ArcGIS Tabulate Intersection tool was used to perform these calculations and produce average values for squares of each place-type.

Current conditions for San José's built environment were assessed using impervious surface cover from National Land Cover Database (2016) and building area based on US Building Footprints from Microsoft (2018). The Zonal Statistics and Tabulate Intersection tools in ArcGIS were used to calculate impervious cover and building area values, respectively, for cells of each urban place-type.

ECOSYSTEM SERVICES

To quantify ecosystem services provided by trees in the current and future scenario model cells, iTree Eco v6.x86 was used in combination with EarthDefine SpatialCover Tree Datasets (2013), a street tree inventory for the City of San José (2018), and allometric equations from the US Forest Service's Urban Tree Database (McPherson et al. 2016).

The Urban Tree Database provides species-specific allometric equations that relate tree age to diameter at breast height (DBH), and DBH to tree height and canopy diameter. These equations were used to project the growth of existing trees and newly planted trees through 2070.

Equations are available for the twenty most common trees growing at sixteen study sites in urban environments across the United States. The allometric equations used in this analysis come from the Forest Service's study site in Berkeley, CA, where available. Where equations were not available from Berkeley, those from the nearest available study site to San José were used.

Current ecosystem services were quantified by first creating three clipped versions of the San José street tree inventory, one for the cells of each urban place-type. Records for species not found in the Urban Tree Database were removed, as were records of vacant wells, trees with heights or diameters of zero, and trees listed as having diameters exceeding 100 inches or heights exceeding 100 feet. Records for the most common twenty remaining species were then isolated. EarthDefine tree points within each Model Place were each randomly assigned to a record (with its associated species, height, and diameter information) within the corresponding clipped version of the street tree inventory. The number of EarthDefine trees for each Model Place exceeded the number of records in the clipped street tree inventories. Thus, many street tree records were assigned to multiple points. The final datasets of tree points and associated size and species information were used as inputs for iTree Eco v6.x86, which produced estimates of the total air pollution removal, carbon storage, carbon sequestration, and avoided runoff achieved by trees in each Model Place.

Future ecosystem services were calculated based on existing tree sizes and species, accounting for tree growth, death, removal, and additional tree plantings. Ages of existing trees in each Model Place were estimated using age-DBH equations from the Urban Tree Database. Their DBHs, total heights, and canopy diameters in 2070 (after fifty years of growth) were then estimated using the same equations. Where the diameter of an existing tree exceeded the maximum value used to model the age-DBH relationship for that species, the maximum age used to produce the model was assigned to the tree. Existing trees were assumed to have a 3% annual mortality rate, likely a low estimate based on various studies in California cities (Roman et al. 2016). For each year, 3% of trees were removed from the existing trees dataset and replaced with five-year-old saplings. A species designation was randomly assigned to each replacement sapling based on the current species distribution for the corresponding urban place-type. Total heights, DBHs, and canopy diameters in 2070 were then calculated for these saplings using Urban Tree Database equations. Finally, additional trees were added to each future-scenario Model Place to illustrate the benefits of greening. The number of new plantings

	Urban Neigh-borhoods	Office Park	Cul-de-Sac Suburb
New Trees	1,300	2,200	1,500
Trees Removed	130	200	70
Net Tree Change	1,170	2,000	1,430
Proportion Oak Woodland	50%	25%	60%
Proportion Riparian Mix	50%	75%	40%
Proportion *Quercus agrifolia*	40%	20%	12%
Proportion *Quercus lobata*	10%	5%	48%
Proportion *Populus fremontii*	10%	15%	8%
Proportion *Alnus rhombifolia*	10%	15%	8%
Proportion *Acer macrophyllum*	10%	15%	8%
Proportion *Platanus racemosa*	10%	15%	8%
Proportion *Aesculus californica*	10%	15%	8%

for each future scenario was estimated based on visual inspection of aerial imagery to identify planting opportunities. New trees were assumed to be native species, either representing coast live oak (*Quercus agrifolia*) woodland or mixed riparian forest habitat types (ICF International 2012). Mixed riparian forest species were added in wetland, riparian, or bioretention areas, whereas oak woodland species were added in drier conditions. The following table reports the total number of trees added and proportion assigned to each species for each Model Place:

New tree plantings were assumed to occur in three waves, with 20% planted in 2030, 60% planted in 2040, and 20% planted in 2060. All newly planted native trees were assumed to be five-year-old saplings that survive through 2070. After calculating DBH, tree height, and canopy diameter in 2070 for trees planted during these waves, iTree Eco v6.x86 was run using these data for all trees in each future-scenario Model Place. The iTree output quantified total air pollution removal, carbon storage, carbon sequestration, and avoided runoff achieved by trees in each future-scenario Model Place.

FUTURE PARK AREA AND CANOPY CONDITIONS

The amount of publicly accessible greenspace in each future-scenario Model Place was estimated by summing the area of existing land listed in the California Protected Areas Database (2020) with additional areas visually identified as well suited for the addition of greenspace. Future canopy cover for each square was calculated by first calculating the canopy area of each future tree using the canopy diameters described in the preceding "Ecosystem Services" section. Canopy areas were summed for each Model Place, then reduced by 25% to account for overlap among trees.

Rural place-types

CURRENT LANDSCAPE CONDITIONS

Current land cover in Coyote Valley's rural place-types was assessed using vegetation mapping from the Santa Clara Valley Habitat Plan (ICF International 2012). The Tabulate Intersection tool in ArcGIS was used to find the land cover breakdown for each rural place-type cell. Current canopy cover in rural place-types was assessed using EarthDefine SpatialCover Tree Data (2013).

CARBON STORAGE QUANTIFICATION

For each rural place-type, total carbon storage for current and future scenarios was estimated by summing storage in three carbon pools: soils, restored riparian habitat, and upland trees.

Soil carbon storage was quantified using soil organic carbon (SOC) values from the USDA NRCS's Gridded Soil Survey Geographic Database (gSSURGO). SOC values from the top 30 cm of soil were used for this analysis, and the ArcGIS Tabulate Intersection tool was used to summarize SOC for each place-type cell. Total SOC from gSSURGO was added to both current and future total carbon storage values for each place-type.

Change in riparian ecosystem carbon storage was estimated using the Carbon in Riparian Ecosystems Estimator for California (CREEC) Tool. In all areas, riparian restoration was assumed to be conducted using planted communities that represent either mixed riparian forest or willow riparian forest communities. Restored mixed riparian forests were projected to contain 75% dominant tree species (*Platanus racemosa, Quercus lobata, Quercus agrifolia, Salix laevigata, Umbellularia californica*) and 25% associated tree species (*Juglans hindsii, Salix spp., Aesculus californica, Populus fremontii, Acer macrophyllum*), in equal proportions. Restored willow groves were projected to contain 90% dominant tree species (*Salix lasiandra, S. laevigata, S. lasiolepis, S. exigua*) and 10% associated tree species (*Alnus rhombifolia, Acer macrophyllum,*

Place-type	Site Preparation	Original Land Use	Planted Community Area (Acres)	Planted Community	Additional Carbon (Tons)	Total Added Riparian Carbon (Tons)
Rural & Open Space	High/Mechanical	Crops	25	Riparian Mix	1573	
Rural & Open Space	High/Mechanical	200	2	Riparian Mix	119	1590
Cultivated Land	High/Mechanical	Crops	4	Riparian Mix	236	
Cultivated Land	Low/Non-Mechanical	Grazing	24	Riparian Mix	1310	1546
Parks & Protected Areas	High/Mechanical	Crops	12	Riparian Mix	707	
Parks & Protected Areas	Low/Non-Mechanical	Grazing	1	Riparian Mix	55	
Parks & Protected Areas	Low/Non-Mechanical	Degraded/Invaded	4	Riparian Mix	223	
Parks & Protected Areas	Low/Non-Mechanical	Grazing	35	Willow Mix	1173	2157

Platanus racemosa, Quercus agrifolia), in equal proportions. Land uses as mapped in the Santa Clara Valley Habitat Plan (ICF International 2012) were used to determine the amount of each land use type to be restored to riparian habitat, and site preparation inputs were based on this original land use. The following table describes the CREEC inputs and total riparian carbon storage output for each rural place-type:

Carbon storage by trees in non-riparian areas was quantified using the area of non-riparian tree canopy in each place-type cell and a canopy-to-storage conversion ratio of 1.365 tons per acre from iTree Canopy. Current tree canopy was based on EarthDefine SpatialCover Tree Data (2013), and summarized for each place-type cell using the ArcGIS Tabulate Intersection tool. For future conditions, additional tree canopy was added along roadsides, on properties, and in restored upland habitats.

SOIL GREENHOUSE GAS SEQUESTRATION & PAYMENT FOR ECOSYSTEM SERVICES

Annual greenhouse gas emission reductions and potential payment for ecosystem services on agricultural lands were quantified using the COMET Planner tool from the USDA's NRCS. The tool was used for agricultural areas in Cultivated Land and Rural & Open Space future scenarios. Input acreages for climate-smart agricultural practices were based on areas of each land use type (ICF International 2012) and visual assessment of opportunity areas using aerial imagery and Rhino 3D.

Appendix 2: Tree Species and Ecosystem Services

Trees along urban and suburban streets provide various benefits to people. Their canopies trap rainfall and help slow urban runoff, and their shade provides shelter from urban heat. They remove pollutants from the air through direct deposition on and uptake through their leaves, and sequester and store carbon dioxide to help regulate the global climate (Silvera Seamans 2013). Different tree species provide these services to differing degrees and require differing amounts of inputs (e.g., water, light, and soil nutrients) to do so. The following charts are intended to inform future tree planting efforts in San José and Coyote Valley by comparing the degrees to which ten tree species provide these ecosystem services and the species' respective water requirements. The four non-native species included are the most common trees currently growing along San José's streets, while the six native species included were common in Silicon Valley's oak and riparian woodland ecosystems historically (Beller et al. 2010).

To quantify ecosystem services for these species, this analysis used allometric equations from the US Forest Service's Urban Tree Database (McPherson et al. 2016), described in Appendix 1, in combination with iTree Eco v6.x86. The allometric equations used in this analysis come from the Forest Service's study site in Berkeley, CA, where available. Where equations were not available from Berkeley, those from the nearest available study site to San José were used. This analysis used tree diameter and height at age 25 as inputs for iTree Eco. For iTree Eco to calculate the cooling benefits of trees, all trees were assumed to be twelve feet to the west of the nearest building. Charts on the following page display the iTree Eco outputs for carbon storage, carbon sequestration, avoided runoff, pollution removal, and energy savings on cooling.

The Water Use Classification of Landscape Species (WUCOLS IV) system classifies landscaping tree species based on their water requirements. WUCOLS classifications vary depending on where a tree is growing in California. The final chart on the following page reports WUCOLS values for each species, assuming they are growing in San José.

Ecosystem Service comparison of San José's four most common street trees (which are all non-native species and are shown in orange) with proposed local native tree species (shown in green).

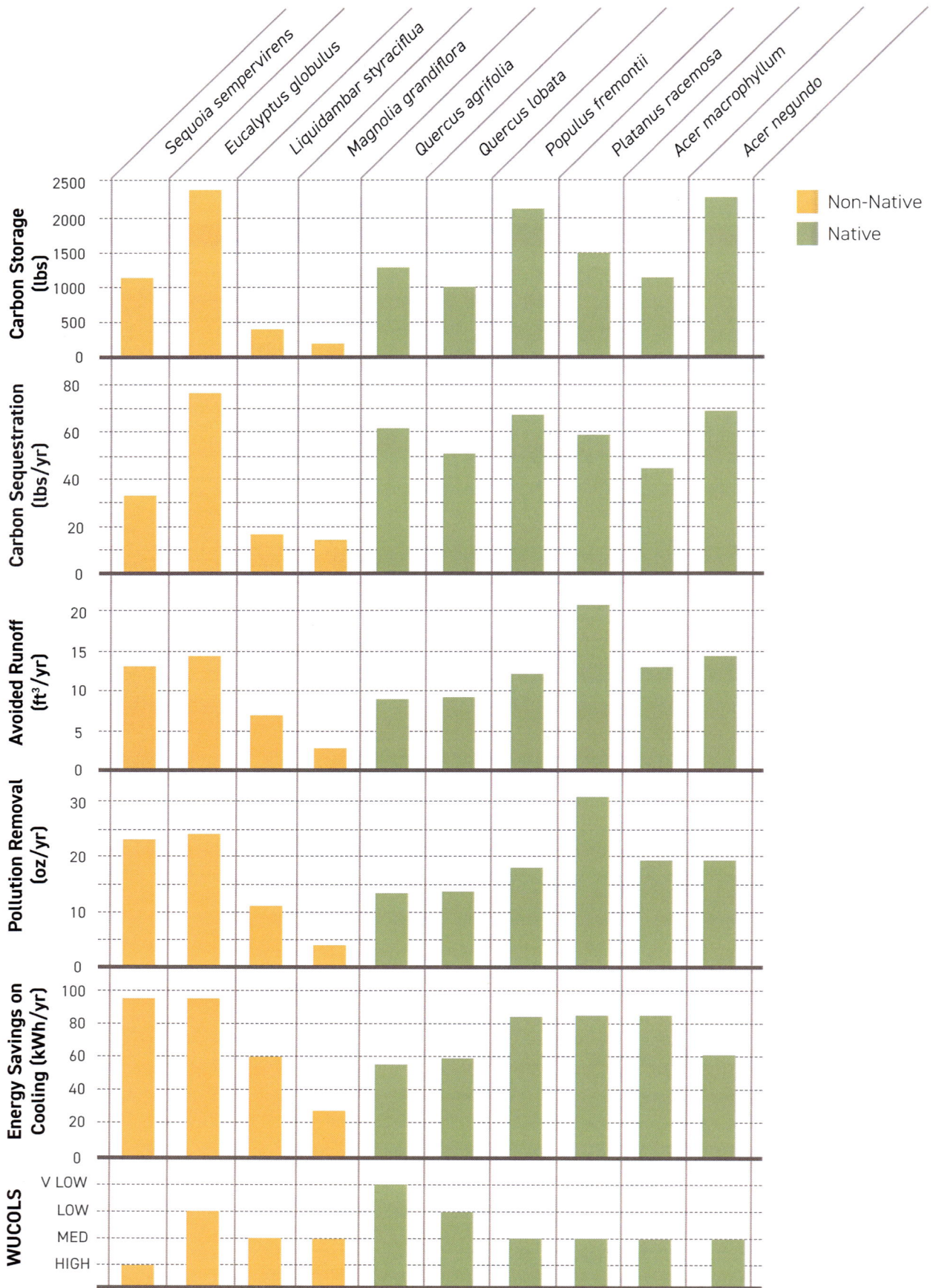

Appendix 3: Methods for Assessing Environmental Disparities

Two analyses were run to assess the equity of park access and tree canopy cover distribution in Santa Clara County. The first analysis examined how canopy cover near one's home varies across income brackets for households of different races and ethnicities. Demographic data for this analysis came from the U.S. Census Bureau's 2017 American Community Survey (US Census Bureau 2017). Data on the number of households within each income bracket were obtained for each of the three most populous races and ethnicities in Santa Clara County: White, Asian, and Hispanic or Latino. These data were available for each census tract in the county. To determine canopy cover near each household, census tract boundaries were clipped to areas zoned as residential according to the Association of Bay Area Governments (2006). The ArcGIS Generate Random Points tool was then used to draw a point in each residential area for every household in each income bracket. Ten-meter buffers were drawn around each household point, and then the ArcGIS Zonal Statistics tool was used to find the average canopy cover value within each buffer. A one-meter-resolution canopy cover raster for urbanized areas of Santa Clara County (EarthDefine 2013) was used for this analysis. Canopy cover values for each buffer were then averaged for each income bracket within each race or ethnicity, and each category was graphed with its standard error. To ensure that random point assignment did not bias results, this analysis was repeated five times with unique random points for each iteration. No significant differences ($p>0.05$ in two-sample T-tests) occurred between average canopy values for each race-income category in the first analysis and those in subsequent iterations. While higher-income households tend to be located within areas of higher canopy across races and ethnicities, the average canopy surrounding White households of any income bracket exceeds the averages of all but the wealthiest bracket of Asian or Hispanic/Latino households.

The second analysis examined how the amount of public greenspace near one's home varies across income brackets for people of different races and ethnicities. The same methodology was used to generate random points for households of each race and income bracket within residential areas of Santa Clara County. One-mile buffers were drawn around each household point, and the ArcGIS Tabulate Intersection tool was used to find the acreage of greenspace within each buffer. The California Protected Areas Database (CPAD 2020) was used to identify greenspaces in this analysis. Park area values for each buffer were then averaged for each income bracket within each race or ethnicity, and each category was graphed with its standard error. To ensure that random point assignment did not bias results, this analysis was repeated five times with unique random points for each iteration. No significant differences ($p>0.05$ in two-sample T-tests) occurred between average park area values for each race-income category in the first analysis and those in subsequent iterations. Across races and ethnicities, higher income households tend to be located in areas with more nearby parkland. Within each income bracket, White households tend to have the greatest acreage of greenspace within one mile.

www.ingramcontent.com/pod-product-compliance
Lightning Source LLC
Chambersburg PA
CBRC091035220326
41597CB00009BA/164

9 781950 313167